John Wolseley

Tracing the Wallace Line

BENDIGO ART GALLERY

7 July - 12 August 2001

Director's Foreword

Tracing the Wallace Line is unique in its installation at Bendigo Art Gallery. In recent times, John Wolseley has travelled and worked in Northern Australia and the Indonesian archipelago, traversing the Wallace Line. Like Alfred Russel Wallace, the founder of biogeography, Wolseley is preoccupied with the flora and fauna of Australia and Asia. Whereas continental drift in the Southern Hemisphere meant that the continents moved apart, in recent geological times the Indonesian archipelago plates have moved closer together, so that separated evolving worlds now collide.

Wolseley has covered the wall surfaces with drawings and placed framed works and museum cabinets within the exhibition space. This installation encourages a dialogue between the hidden and the revealed in both the symbolic wall drawings and the detailed work on paper.

Bendigo Art Gallery is proud to host *Tracing the Wallace Line*. An exhibition of this nature relies heavily on the support of certain individuals and organisations. I would like to thank the curatorial team who have assisted with the development of this exhibition. Paul Carter eagerly approached the task of writing about John's work and we are grateful for his catalogue essay. The enormous job of assisting John with his wall drawings would not have been possible without the support of Ian Ellis and Max Ellis Signs. I would also like to acknowledge especially John Wolseley, who has made the exhibition process both enjoyable and rewarding through his tireless enthusiasm and commitment to the project.

Karen Quinlan

Tracing the Wallace Line

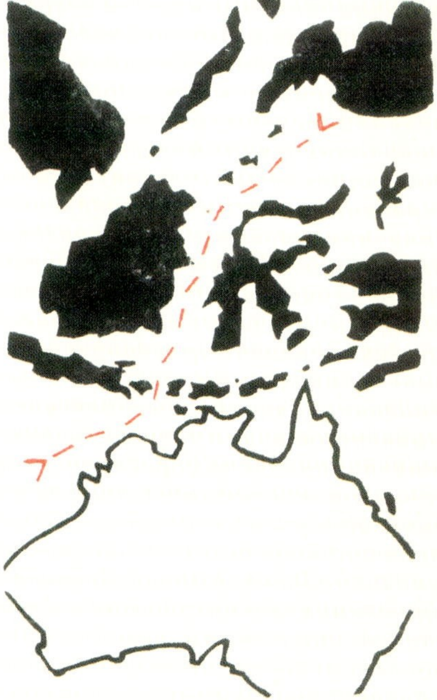

The Australo/Papuan continental plates (once Gondwana) and the plates which made up Laurasia were far apart when many of the animal bird and plant species evolved. In the north - pheasants, tigers, monkeys hoopoes and magnolias. In the south - honeyeaters, echidnas, tree kangaroos and myrtle beech trees. These rafts of species slowly moved together and in recent geological time they collided to the north of Australia. All along a line - known as the Wallace Line - between Bali and Lombok, and up between Borneo and Sulawesi there is a juxtaposition of differently evolved life forms - different yet often strangely similar owing to parallel evolution. This was first described by the great scientist, Alfred Russel Wallace.

Over the last four years, I have been moving back and forth between the various islands and continents on each side of the line documenting the differences and correspondences I have found. I have set these drawings and paintings within large scale drawings on the gallery wall which represent the abstract dimensions of geology, and the great movements of the earth's dynamic systems.

John Wolseley

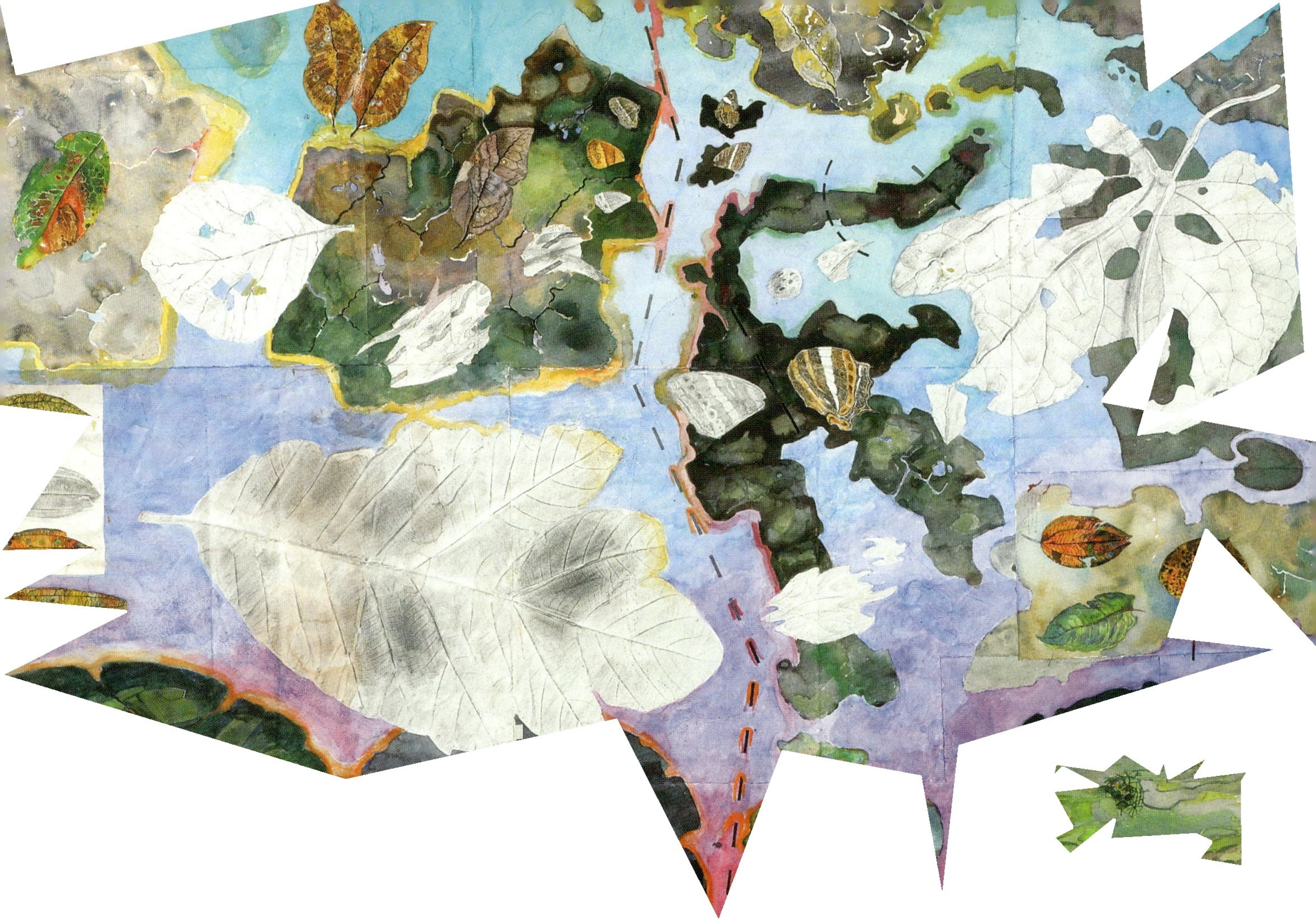

Painting the Scene that Disappears

Paul Carter

In 1996 John Wolseley exhibited *Patagonia to Tasmania, Origin, Movement, Species, Tracing the Southern Continents*. He invited me to write a text to accompany that exhibition. My response, 'The Anxiety of Clearings', contained the following remark: "The forms of the mental universe may bear no resemblance to those of the physical world, but their internal coherence is such that they mimic the external world as if they were exact copies." 'Painting the Scene that Disappears' continues that meditation. PC

Opposite: *Tracing the Wallace Line: wing leaf and land* (detail) 1999 (trace No. 6)

1.

Mark the line.[1] The essential division. Nothing in itself, it makes a difference.

Either side of it waves surge angrily. Avatars of the new void, they face each other spectrally across the line. A world of doubles and halves is born. They will go to war to prove their priority.

Butterflies are airborne symmetries. They carry their mirrors with them. They wear their bifurcations gracefully. When they close their wings their doubles disappear. In the eye of the sun a dead leaf.[2]

The principal drawback [of Kay island] for a collector is the want of good paths, and the dreadfully rugged character of the surface, requiring the attention to be so continually directed to securing a footing as to make it difficult to capture active winged things, which pass out of reach while one is glancing to see that the next step may not plunge one into a chasm or over a precipice.

Butterflies do not fly in straight lines. You cannot plunge them into chasms. Even Stalin could not make them fly faster. Their winking, wambling path suggests an obstacle course invisible to the eye.

Creatures of chaos, they see creation everywhere and stagger in its glare.

The lives of butterflies have a pattern of their own.

The line is the grief of tectonic plates. A shuddering impotence. Out of it tsunami spring, oppositely pairing off. They stumble, coastlines spring up. The gulf between their broken halves is hungry for ships.

On either shore wrack, especially of the genus Fucus, is deposited. Also Walking Leaves, Stick Insects. And Lepidopterous wings (*Kallima inachis*). Or, turned over: *Kallima paralekta*.[3]

1 In 1858 Wallace reported, "In the Archipelago there are two distinct faunas rigidly circumscribed, which differ as much as those of South America and Africa and more than those of Europe and North America. Yet there is nothing on the map or on the face of the islands to mark their limits. The boundary often passes between islands closer than others in the same group." In 1863, "He drew a red line on the map passing down the Makassar Strait. To the west he wrote 'Indo-Malayan region' and to the east he wrote 'Australo-Malayan' region'. This became the Wallace Line." (Penny van Oosterzee, *Where Worlds Collide*, Reed, 1997, 34.)

2 "Marvellous to relate, it does possess the habit of almost invariably entering a bush loaded with dead leaves, and is so instantly lost to sight." (A.R. Wallace, 'The Disguises of Insects', *Studies, Scientific and Social*, Macmillan, 1900, vol 1, 193.)

3 A.R. Wallace, 'The Disguises of Insects', *Studies, Scientific and Social*, vol 1, 191. The Malayan K. paralekta is an "ally" of the "common Indian butterfly, Kallima inachis."

Opposite: *Tracing the Wallace Line: wing leaf and land* (detail) 1999 (trace No. 6)

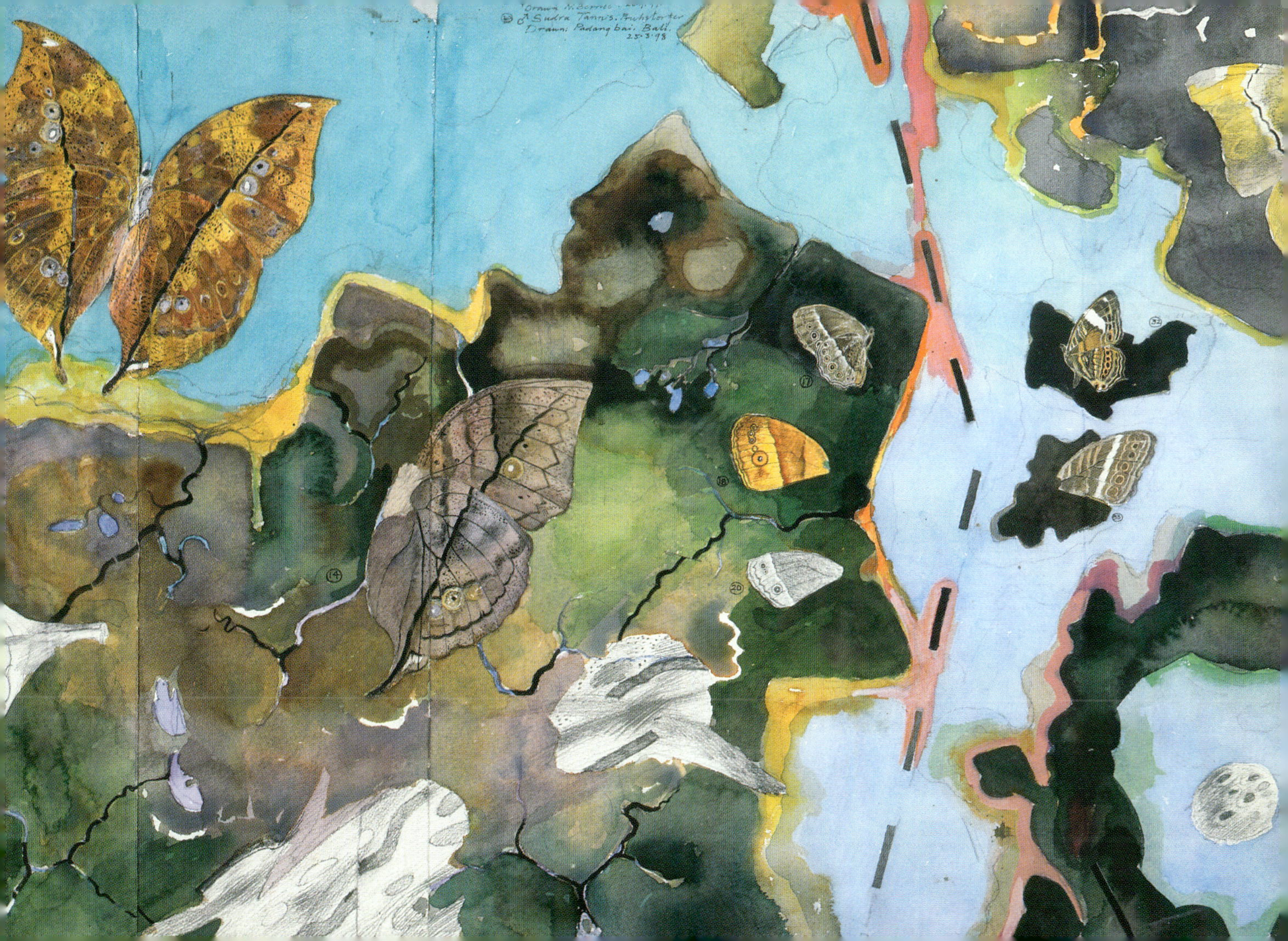
Sudra Tannis.
Drawn: Padangbai. Bali.
25.3.98
14
17
18
20

2.

Every word, every soul-breath, mimics a world.

Malaysian authorities have had to extend the state of emergency in Sarawak for the sixth day - shutting down schools, businesses and factories and ordering some two million people to stay indoors. Poor visibility caused by haze ...[4]

The words *breath* and *life* imitate "the life-giving action of breathing." *Growth* expresses "the gradual extension of a young plant." *Mountains* are named after "the natural panting ejaculations of those who climb them." Pronouncing "the word *fire* we seem to imitate with the lips and breath the wavy flickering motion of flame."[5]

Forest fires in drought-hit parts of Indonesia have so filled the air with smoke and soot over a vast area ...

Cough, wheeze, closely represent in words actions ...

... from Sumatra to Singapore, from Borneo to Brunei and from the Malaysian province of Sarawak to the Indonesian island of Sulawesi that people are gasping for breath.

Greeks called the nightmare after the giant *Ephialtes*: he jumps up to the heavens "and will not permit the Gods to abide there." *Incubo*, the Roman name, means asphyxiator, he who weighs down the sleeper, squeezing out the life-breath.[6]

Situation aggravated by timber mafia keen to make the most of conditions, and farmers, particularly in Java, clearing land for cultivation by burning ...

"I seemed first to hear a distant beating or flapping sound, as of some creature with huge wings, the sound came nearer and nearer, till at last a deep thud was heard and the flapping ceased."

"In *Epaphos* we should see an animal demon, the HOOPOE bird."

"I then seemed to feel that the creature was clinging with its wings outspread against the wall of the house just outside my window, and I waited in a kind of fearful expectation that it would come inside."[7]

The flame-winged *papuscia*[8] or butterfly-bird, in Latin *papilio*. Nightmares have their colours too.

4 *Times of India*, 25 September, 1997 for this and related quotations.

5 A.R. Wallace, 'The Expressiveness of Speech', *Studies, Scientific and Social*, vol.2, 115-137, for this and related passages.

6 W.H. Roscher, 'Ephialtes: A Pathological-Mythological treatise on the Nightmare in Classical Antiquity', in *Pan and the Nightmare*, Spring Publications, 1974, 46, for this and later quotation

7 A.R. Wallace, *My Life*, Bell, 1905, vol 1, 27. An "often-repeated" childhood dream and, Wallace writes, "not nearly so distressing as many of the nightmares I have had since."

8 Mario Tortorella, *Dizionario Dialettale dell'Avifauna Leccese*, Congedo Editore, 1981, 66.

3.

INDANG INDANG[9]

At intervals they utter a peculiar cry of two notes. My practice was, therefore, on detecting any sign of a Pitta's vicinity, to stand motionless and give a gentle whistle occasionally, imitating the notes as near as possible.[10]

"The method of catching wild pigeons ... brings the animistic ideas of the Malays into strong relief, the 'souls' of the wild pigeons being repeatedly referred to."[11]

The conception of such a double personality in each of us, a second-self, which in most cases remains unknown to us all our lives[12].

"First you build a small sugar-loaf hut called *bumbun* ... before entering the hut the wizard must go through the 'Neutralising Rice-paste (*tepong tawar*) ceremony."

After half an hour's waiting, I was often rewarded by seeing the pretty bird hopping along in the thicket.

Said to live an independent mental life, to have means of acquiring knowledge our normal self does not possess, to exhibit all the characteristics of a distinct individuality with a different character from our own.

"After a time the decoy-bird's challenge is met by first one and then many counter challenges, then the wild pigeon approach."

Then I would perhaps lose sight of it again.

Is surely a conception more ponderously difficult, more truly supernatural than that of a spirit world.

My gun raised and ready for a shot, a second glimpse would enable me to secure my prize.

"The 'call' will occasionally attract to the spot wild animals such as deer and tigers. Is it not possible that the story of the lute of Orpheus may have had its origin in some old hunting custom of the kind?"

Praise the artist who, observing correspondences, does not condemn what he draws to death.

WAK-WAK.

9 This and "wak-wak" are opening phrases of Malay pigeon-catching charms. (W.W. Skeat, *Malay Magic*, Macmillan, 1900, 595 & 598. Doubling up everyday speech, spells are a kind of "double Dutch" addressed to spirits.

10 A. R. Wallace in Lombok (Penny van Oesterzee, *Where Worlds Collide*, 103

11 W.W. Skeat, *Malay Magic*, 132.

12 A.R. Wallace. *My Life*, vol 2, 294. Wallace was a keen student of Spiritualism.

Overleaf: *Tracing the Wallace Line: descent of dipterocarp* 1999 (trace No. 10)

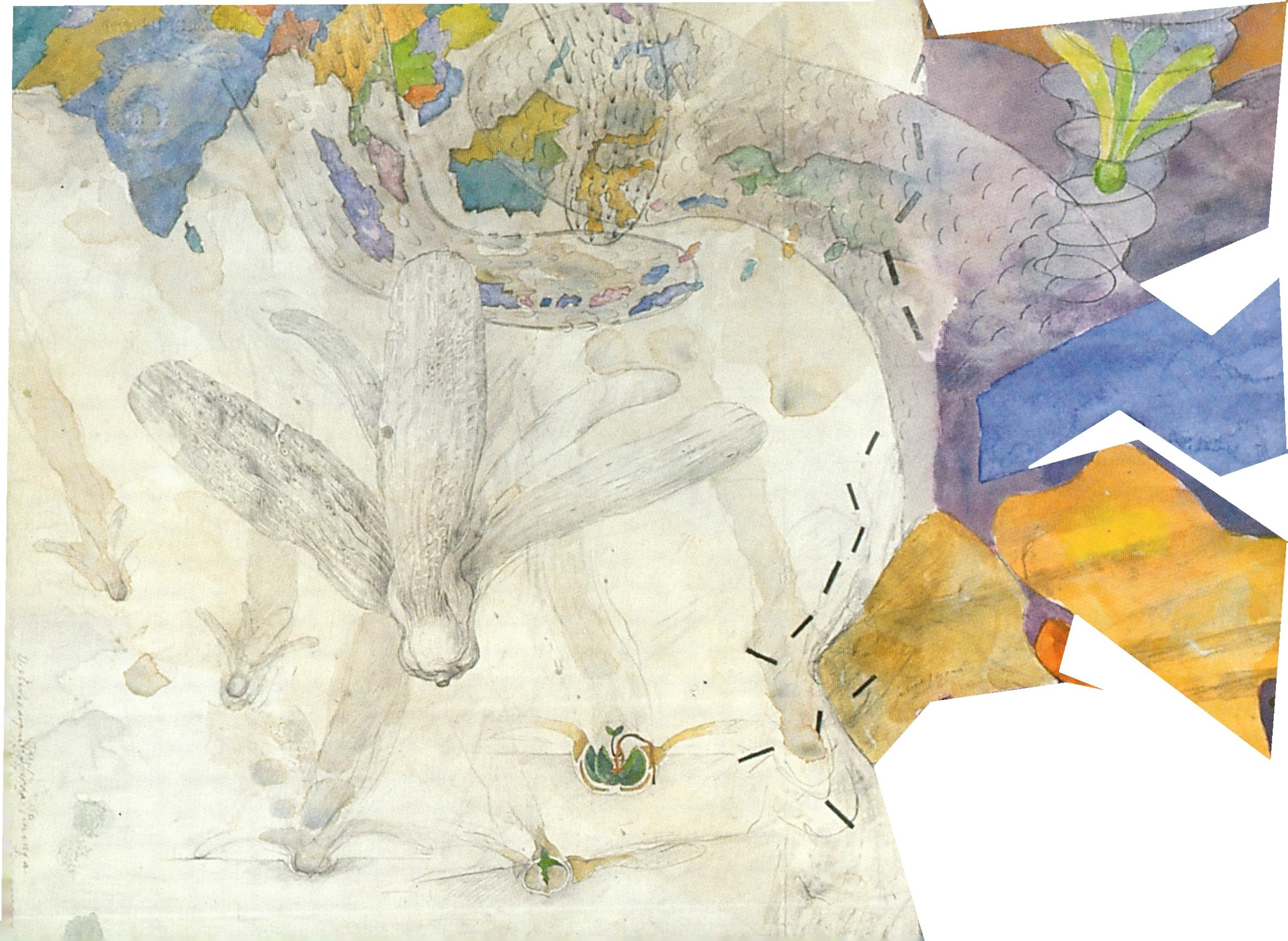

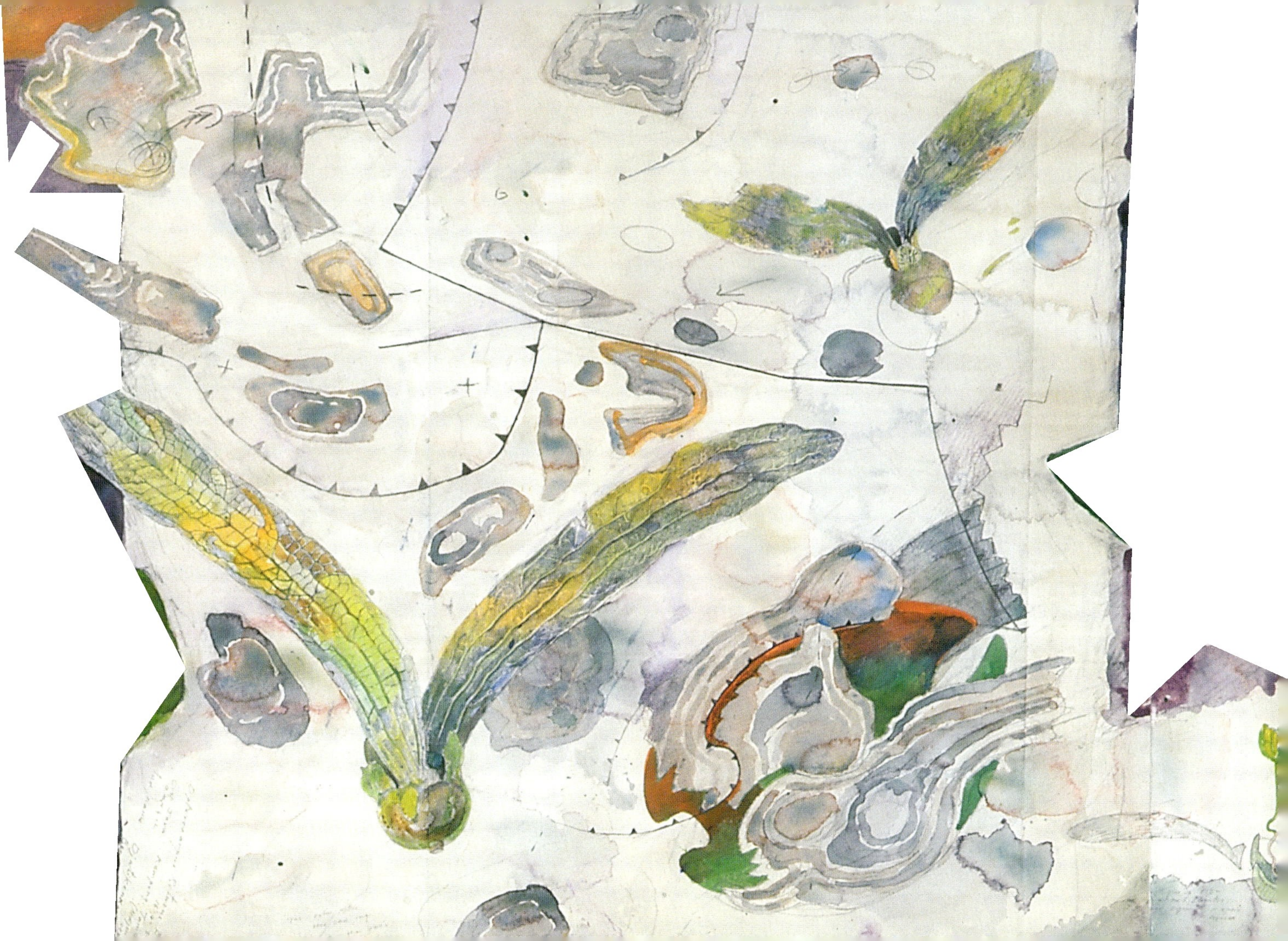

4.

I had considerable mesmeric power. [So did] my brother Herbert ... when we were residing together at Manaos, he used to call up little Indian boys, give them a copper, and by a little gazing and a few passes send them into the trance state, and then produce all the curious phenomena of catalepsy, loss of sensation."[13]

A man wrote a treatise *Against Focus*. He wanted the bird's-eye wings, the pearl-scaled clouds, the cobwebbed shafts of silver light, the topography of jewelled creatures to melt into the sound of water.

"The beauty and brilliancy of this insect [the 'bird-winged butterfly'] are indescribable and none but the naturalist can understand the intense excitement I experienced when I at length captured it."[14]

Nightmares assailed him in his sleep. He heard the thunder's cataract as if the canopy were paper tearing. Out of its fiery fissure red birds of paradise streaming. Ants defile across his brow.

[Red birds of paradise] were in constant motion, jumping about the cage from perch to perch, clinging on the top and sides, and rarely resting a moment the first day till nightfall[15].

How these doubles long for partners different from themselves. How they desire to lose the way, but wherever they look scimitars shake the foliaged-tops, and corduroy tracks puddled like well-tempered swords glimmer, and daytime moths like shrunken angels wings besmeared struggle etherised on the surveyor's table.

"On taking it out of my net and opening the glorious wings, my heart began to beat violently, the blood rushed to my head, and I felt much more like fainting than I have done when in apprehension of immediate death."

He smokes to wile away the night. The candle quivers, the mosquito curtains billow. The shadows caught in their folds stretch and contract. Somewhere in-between, islands and continents hybrid regions undulate.

Mosquitos and crickets chain-saw the night.

The second day they would be dull, they were seized with convulsions and fell off the perch. The third day they are almost always found dead at the bottom of the cage.

Too much gazing, sir, too much gazing. It is also the eye that sleeps that sees.

13 A.R. Wallace, *My Life*, vol 2, 275.

14 A.R. Wallace, *The Malay Archipelago*, 342.

15 A.R. Wallace, *The Malay Archipelago*, 537

"LOOK!"[16]

Between the parts of opake and colour'd bodies are many Spaces, either empty, or replenish'd with Mediums of other Densities ... in whose interstitial zones slender Arcs of Colours shaped almost like the Conchoid grow, violet, blue, which, bended more and more about the transparent spot are completed into Circles or Rings.[17]

blackbandsspotsorlargepatchesofbrilliantcolourspaleorgoldenyellowrichcrimsonsorgorgeousmetallicbluesandgr eensthicklyspeckledwithgoldengreendotsintensemetallicgreenorazurebluesimplyblackandwhitepatternsmanyve rystrikingcrimsonorgoldenpatchesatcertainanglesdifferentopalescentthicklyspeckledwithgoldengreendotshues

"We saw there a faint white patch on his coat on the left side. This grew brighter, then seemed to flicker, and extend both upwards and downwards, till very gradually it formed a cloudy pillar extending from his shoulder to his feet."

Geo-gestures need not crush, pummel, bruise. Clapping hands, they produce the Himalayas.

"Again [he] said to us: LOOK. And then clapped his hands."

Parting, they linger awhile. The spindrift capes: clinging fingertips.

"The figure put out her hands, clapped them as he had done, and we all distinctly heard her claps following his, but fainter."

Compressing two Prisms hard together [] I found the place in which they touched to become absolutely transparent.

LAID-OUT-ON-LONG-TABLES-COVERED-WITH-WHITE-PAPER parrots pigeons paradise birds pittas rivalling the Golden Apples of the Hesperides flycatchers drongos blue-billed gapers called by the Malays the "rain-bird" woodpeckers all peculiar barbets cuckoos trogons ...O, this mode of exhibiting bird skins ESPECIALLY SUITABLE FOR ARTIFICIAL LIGHT![18]

"The figure then moved slowly back to him, grew fainter and shorter, and was apparently absorbed into his body."

"After weeks of study, a whole series of specimens, which seemed at first hardly distinguishable, are gradually separated into well-defined species, and ORDER ARISES OUT OF CHAOS.

16 A.R. Wallace, *My Life*, vol 2, 330.

17 Isaac Newton, *Opticks*, Dover, 1952, 195 & 249. Newton's characterisation of colours as impurities identifies clarity of vision with transparency. Likewise, in the spectrum of Wallace's classification, the fascinating brilliance of nature is inevitably lost. I associate this epistemological blindspot with a propensity for seeing ethereal bodies where there were none.

18 A.R. Wallace, *The Malay Archipelago*,

6.

"A green catbird yesterday was foraging in its own world quite high in a tree."[19]

Others, differentiating themselves from others, live in a simple world. By contrast the mimic imagines *another* other. If there was not a third position, a scientific pair of eyes, dedicated to his discovery, he could dispense with his disguise.

"I was thinking about the correspondences between this catbird and a bird occupying a similar niche on the other side of the Wallace Line ... gesture, manner of flight and diet."

Splitting the world in two, the line inaugurates the drama of translation. Birds, insects, plants are hieroglyphs of a forgotten language. Partially exposed in the cut of the eye, their display is the injury of their exposure. Previously, in the dark-light corridors of various vegetable heights, they had enjoyed a birth without edges (their habits of nidification unknown).

"By moving one's vision from one bird to another of a completely different species, looking at one and then the other with an intense scrutiny and back again, one might really see the bird or beyond the bird."

The artist, neither scientist nor scéantist, who sees through this must incubate a vision without scale, sight ways without a vanishing point. He scales the rigging of lianas, the turning kite of his eye notches the sky.

The better his mimicry, the less he represents. Pressed up against the surface, in his spiritual prism there is caught a finely-veined wing of transparency; a world-window without frames. It belongs to the third position.

"After dwelling so long on this mysterious axis of difference ... these paintings could create a compounding or intensification."

The third position is the general principle of repair.[20] The scientist identifies differences. The scéantist deals in doubles. The artist is the observer of these dividing, recombining destinies. He sees the different habitats superimposed. He hears the different sonic spectra together. He is the collector of interference patterns, the connoisseur of fringes.

He copies the forces acting in any region of space. Master of his own medium, he is sympathetic to the reciprocal action of body-mass and wingbeat. He presumes an elastic air, comparable with the loaded brush's coefficient of drag.

It is to this perspective that he owes the intuition of analogous forms and growths. The convergent evolution he detects is not measured in continental drift but in the strangely familiar upsweep of an unidentified bird, and its shadow's volume striping the trunk.

19 John Wolseley, adapted from 'Lamington National Park, 17 September 1999, notebook.

20 See D'Arcy Wentworth Thompson, *On Growth and Form*, Cambridge, 1942, vol 1, 270-271.

Opposite: *The territories of birds remembering Gondwana and Laurasia* (detail) 2001 (trace No. 5)

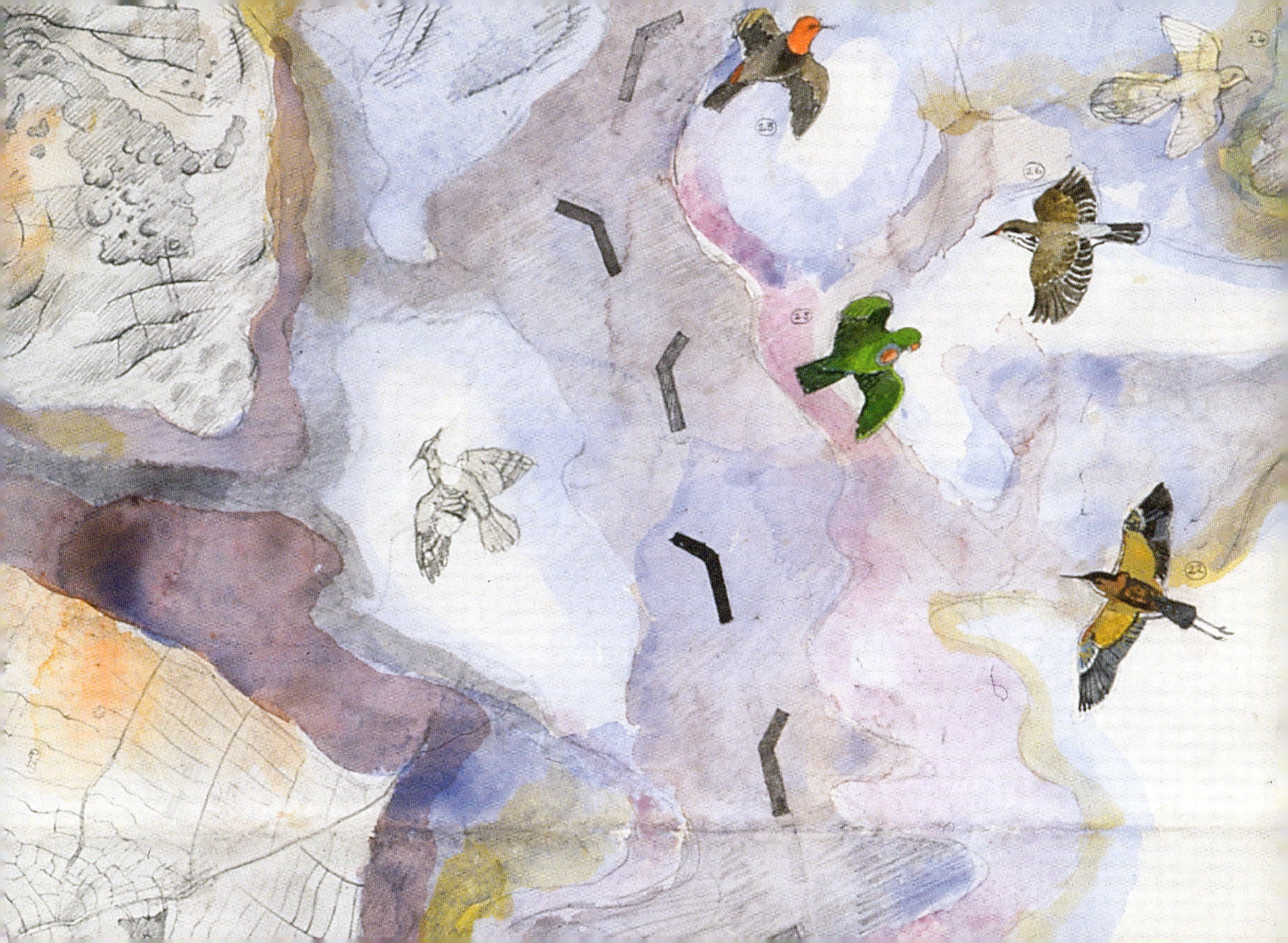
26

7.

Thinking how much this project seems to be about the branching of things.[21]

"[The Sumatran monkeys] take tremendous leaps from the branches of one tree to another a little lower."[22]

THE INDO-MALAY ISLANDS MALAY PENINSULA, SINGAPORE, BORNEO, JAVA AND SUMATRA we have indications of a vast continent.[23]

Not splitting, splayed. Not radiating but remembering what was contracted when the sky climbed out of its egg.

"It is very amusing when one strong leader takes a bold leap, to see the others following with more or less trepidation."

THE TIMOR GROUP TIMOR FLORES, SUMBAWA AND LOMBOK.

The shore is not a division. Transfusing sea and land, where rivers meet their maker, and the rejected sea swings in the hammock of its enemy.

"One or two of the last seem quite unable to make up their minds to leap till the rest are disappearing."

THE CELEBES GROUP SULAWESI AND BUTON gradually, irregularly broken up.

Hail Mambang Tali Harus, spirit of the midway, Si Raya, surfer of the tiderip, mid-current man, by extension master of the gap. Or you, Madu-Raya, whose abode is the navel of the seas, who cause the tide's ebb and flow.[24]

"In desperation at being left alone, they throw themselves frantically into the air."

THE MOLUCCAN GROUP BURU, SERAM, BACAN, HALMAHERA, TERNATE, TIDORE, MAKIAN, KAYOA, AMBON, BANDA, GORONG AND WATUBELA the island of Celebes probably marking its furthest westward extension.

You masters of the double line, put out the false brightness.

"They often go crashing through the slender branches and fall to the ground."

THE PAPUAN GROUP NEW GUINEA, ARU, MISOOL, SALAWATI AND WAIGEO beyond which a wide ocean.

21 John Wolseley, notebook entry, 12 February 1998, Abu's Pensione, Bogor, Java.

22 A.R. Wallace, *The Malay Archipelago*, 143.

23 A.R. Wallace, *The Malay Archipelago*, adapted from the Introduction.

24 W.W. Skeat, *Malay Magic*, 85 & 91-92. Also K.M. Endicott, *An Analysis of Malay Magic*, Oxford, 1970, 103-104. Skeat reports that "When there is a decided glow at sunset, any one who sees it takes water into his mouth and dislodges it in the direction of the brightness." Endicott's structuralist analysis of the divisions of the Malay world wholly prevents him from understanding the doubled, branching genius of these chiasmatic realms.

8.

The destructive character is even handed. He is not angry. He feels no pity. He adheres to the facts.[25]

YA YA TUAN; ORANG-UTAN, TUAN. Close by, sir, on the path to the mines. He can't get away.[26]

The destructive character goes hunting.

I instantly fired. Running, climbing, thick with hanging and twisted creepers. We could see one leg hanging down, useless, broken by a ball. He now fixed himself in a fork: if night falls, he will remain and die in that position. Cut off the branch, cowards! Fired again; he then moved off, to some branches where he could not fall, and lay all in a heap as if dead or dying.

When the destructive man calls, men come running. The ire of the destructive character is aroused by all attachments. They are obstacles to progress, like the night. Bearer of news, he carries a torch. The destructive character blames El Niño. He does not take responsibility.

Frantically, we began pulling at the creepers. We shook the tree very much. Down the giant came with a crash; as large as a man, very like one. To my astonishment. Dreadfully wounded, both legs broken, one hip-joint and the root of the spine completely shattered.

He is free of sensuality. Although he has a good ear for languages, he prefers the collection of bones. And pleasure.

Two bullets in his neck and jaws. It appeared to be a different kind from any I had yet seen. Still alive when he fell. For several minutes he lay groaning and panting. They soon destroyed him with their spears and choppers. Carried him home tied to a pole.

The destructive character sleeps peacefully at night. Sirens pass through his dreams and he does not hear them. Nine hundred forest fires flicker across his eyelids, and the destructive character smiles. When the thunder rolls round the mountains, his closed eyelids flicker like poker machines.

Spent the whole of the next day preparing the skin and boiling the bones to make a perfect skeleton, now preserved in the Museum at Derby.

25 Walter Benjamin ('The Destructive Character' in *One Way Street and Other Writings*, translated by E. Jephcott & K. Shorter, Verso, 1979, 157-159) invented this character. These lines are free variations in the spirit of the original.

26 A.R. Wallace describes a number of orang-utan hunts (*The Malay Archipelago*, 51-66). Penny van Oosterzee excuses the obscenity of these passages on the grounds that Wallace was a man of his times. (*Where Worlds Collide*, 180) That will hardly do. Wallace was also an extraordinary naturalist, a social reformer and a man of empathy and imagination. It is my thesis that the ghosts of his bad conscience came back to haunt him through the medium of the séance. The intensity of Wallace's descriptions of chloroformed butterflies and stuffed birds suggests one in the grip of hallucinations. His hyper-vision rendered him blind to the destruction he wreaked.

Journal Extracts
Crossing the Line 1998-2001
John Wolseley

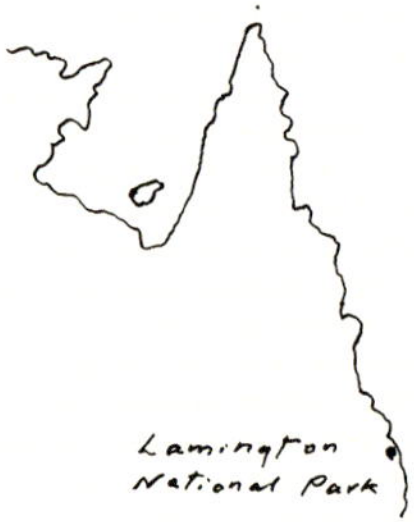

LAMINGTON NATIONAL PARK, 17 SEPTEMBER 1999.

Been drawing the different spaces and habitats within these huge forest trees. Drawing the mid-storey where the first thick branches open out and create their own series of halls and pavilions, and then push themselves further upwards where there are jungly conservatories of crows nest ferns, epiphytes and orchids. I want to show how the different birds inhabit their own levels or spheres. A green catbird yesterday was foraging in its own world quite high in a tree. I watched it hop along a fern-sprouting branch, making that strange child crying noise, and then seeing something on a smaller branch jump onto it and suspend itself upside down . Through the same aerial space a regent bower bird whizzed through - zig zagging this way and that and suddenly up into another green pavilion and out of sight in seconds.

Opposite: *Bower bird, the space between trees* 1999 (trace No. 3)

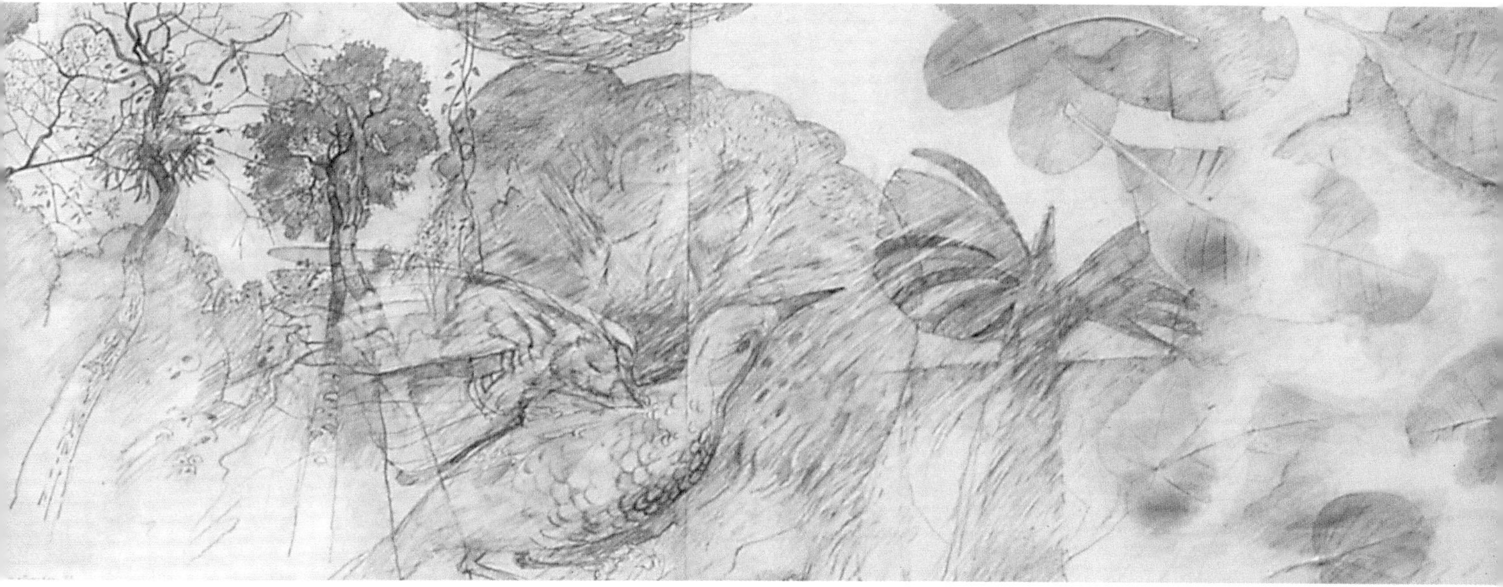

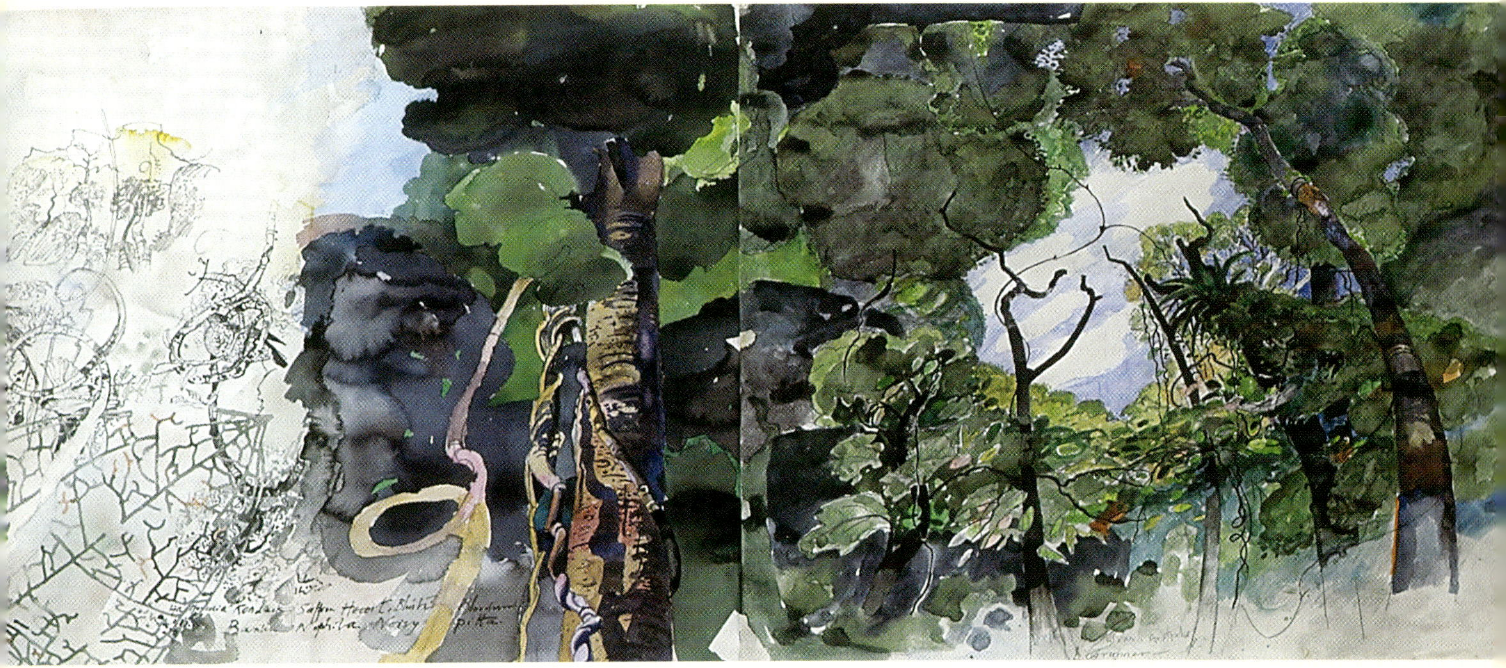

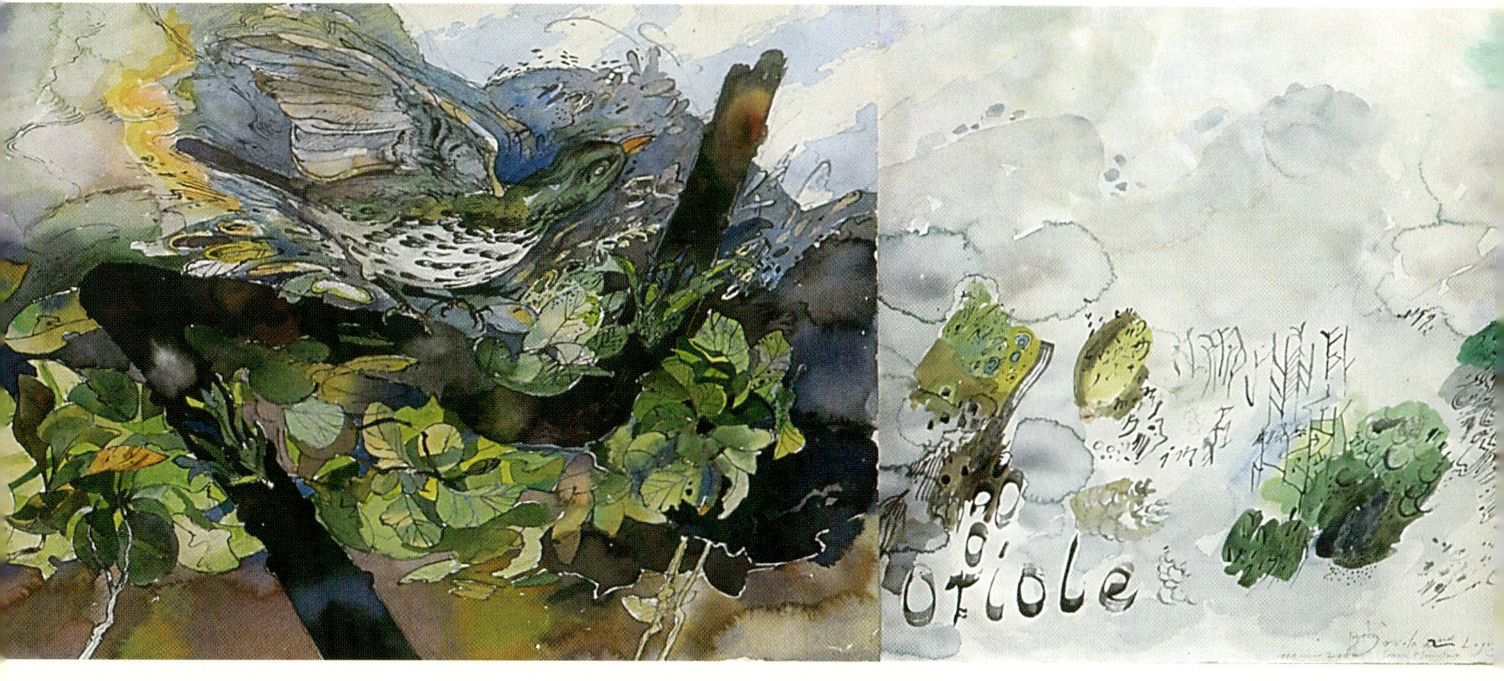
oriole

I was thinking about the correspondences between this catbird and a bird occupying a similar niche on the other side of the Wallace Line -perhaps another fruit eater like a barbet. If I close my eyes and imagine that golden-naped barbet I saw in a dipterocarp forest in Sarawak I can do a kind of film in my head, phasing back and forth between catbird and barbet . There are correspondences of gesture, manner of flight and diet between the essential character of the two birds.

I am not quite sure why I am so fixated on these sorts of things. Perhaps it's some idea that by moving one's vision from one bird to another of a completely different species - looking at one and then the other with an intense scrutiny and back again - one might as a Tertium Quid gain some insight - *really* see the bird or beyond the bird. Rather on the lines of the Buddhist idea that if you truely see a cicada maybe then you could achieve satori or enlightenment. Or at least - being a bit less ambitious about this - by looking at the subtle differences where birds

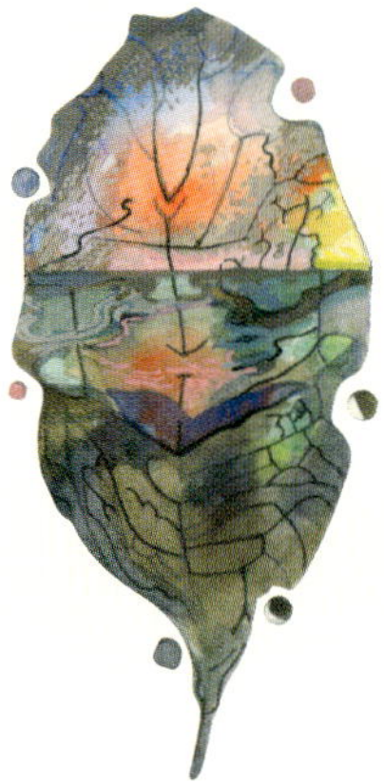

Previous Page: *Rainforest canopy - olive backed oriole* 1999-2000 (trace No. 2)

which are completely unrelated seem almost to have mimicked the other, one gains a heightened awareness of the different characters which make up the bird. For instance if you pair a white-fronted tree creeper with an orange-backed woodpecker you notice the differences in the beaks both used to procure food from the bark of trees - one with a Laurasian method (hammering) and one with a Gondwanan (probing).

I've provisionally called these paintings 'Bird Correspondences of Wallacea' or some such - It seems to me that these paintings I am doing - by setting up a dialogue between the nature of a number of birds and their comparative type across 'the line' - could create a compounding or intensification of these insights. I can't quite articulate it in words yet but I seem to be following some kind of symbolic journey which uses this theoretical line as a pivot or axis on which to rest these meditations.

Overleaf: *The territories of birds remembering Gondwana and Laurasia* 2001 (trace No. 5)

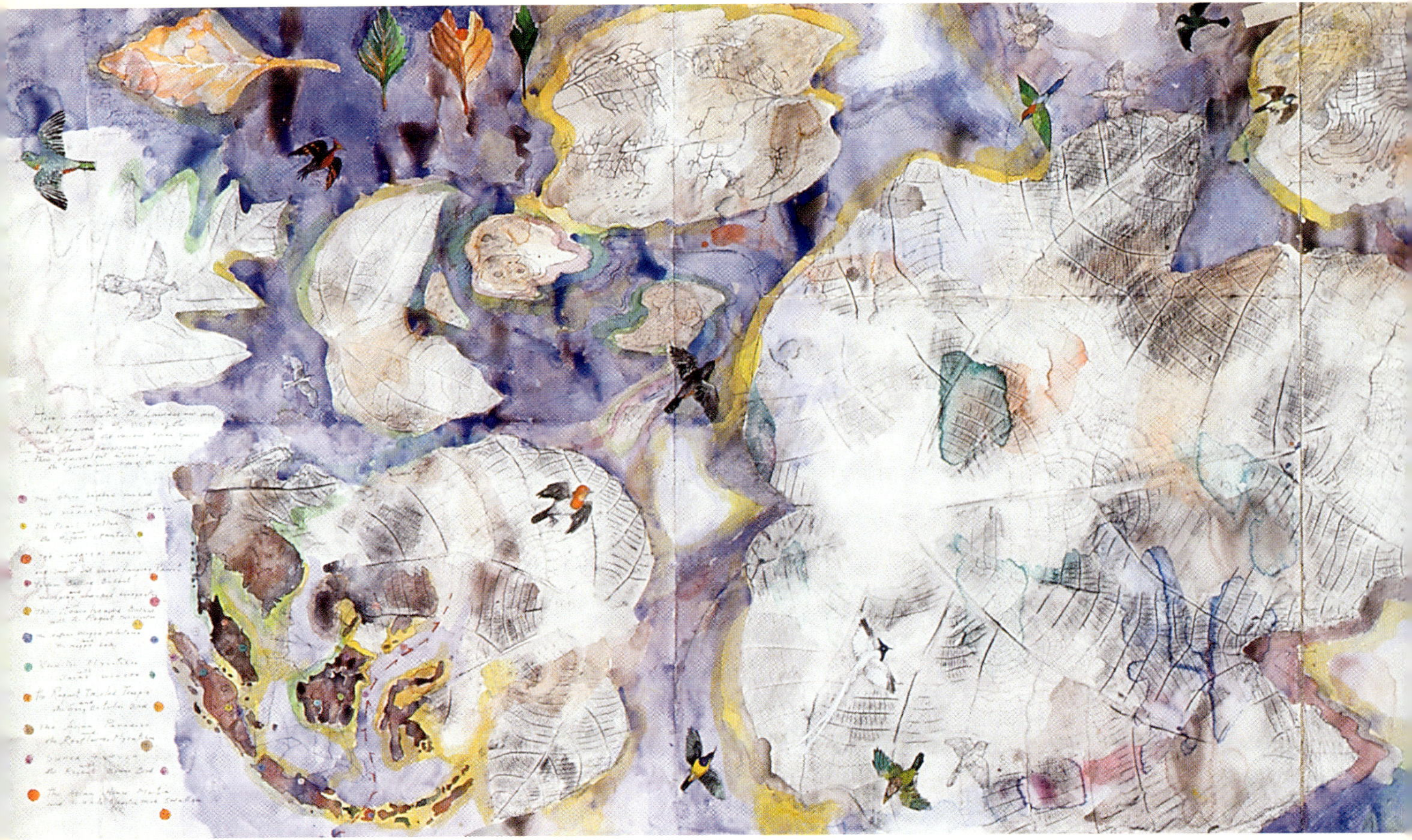

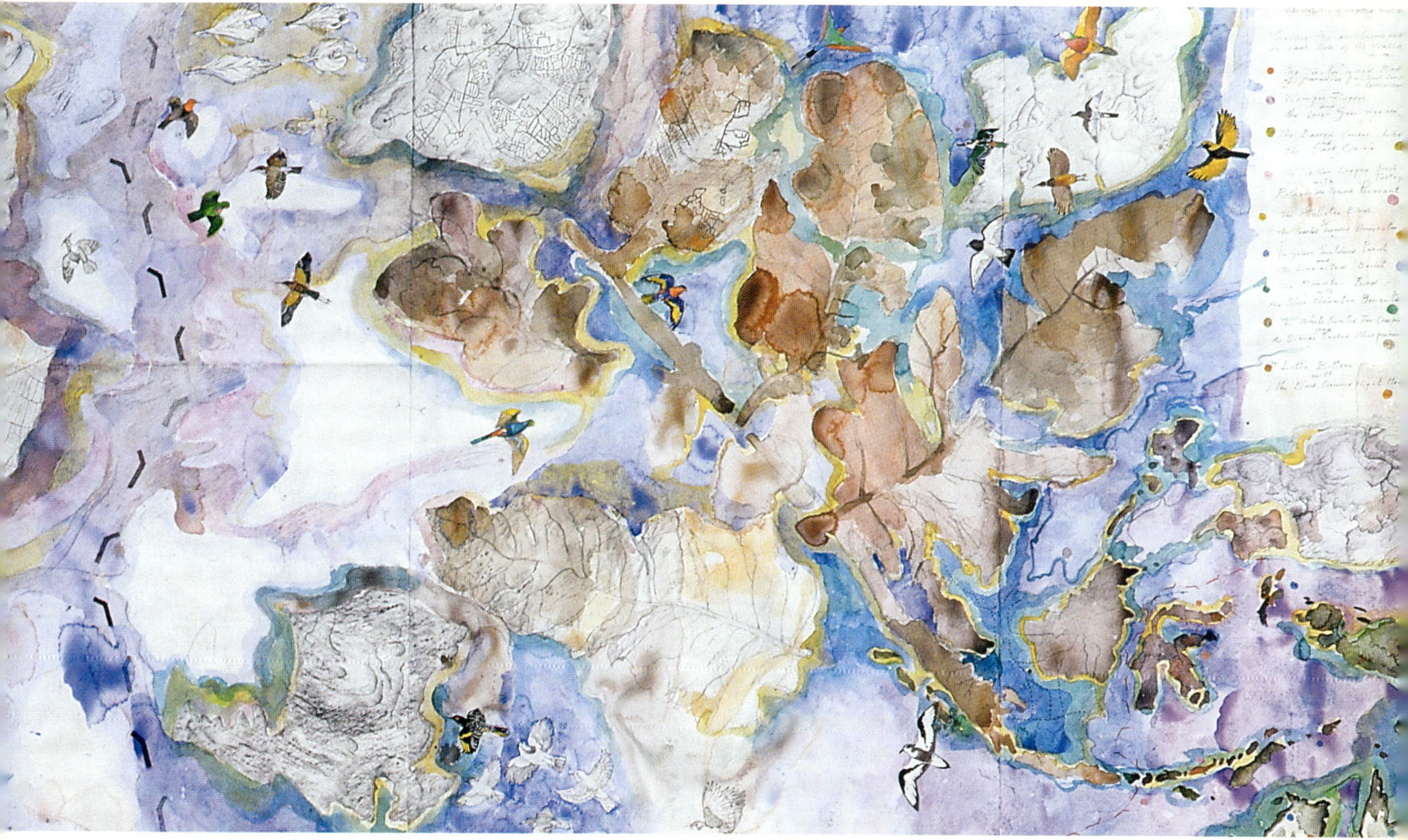

The Wallace line for me, as a continual point of return, seems to be a back-bone - or perhaps a sacrum - which because I continually return to it with an eidetic image of comparative bird, glider, feather or leaf, is becoming a central symbol or icon on which to meditate about the mystery of the unfolding evolutionary power of the universe. ' What is Being?'- or those sorts of hubristic questions which seem to attract old hermits in rainforests.

If our way of understanding the world is often a comparative process of relating one thing to another this particular site of investigation is very engaging. Somehow it includes a chain of metamorphosis all the way up from Bali to North Borneo - a line or axis, the very pivot where one thing becomes another. I remember how when you cross the Wallace Line the continuous sounds of barbets (that tinsmith clinking) fades away and suddenly you are aware that their place has been taken by the sounds of parrots, forests full of parrots and cockatoos.

Opposite: *The architecture of leaves in Java and Malaysia* (detail) 1999 (trace No. 11)

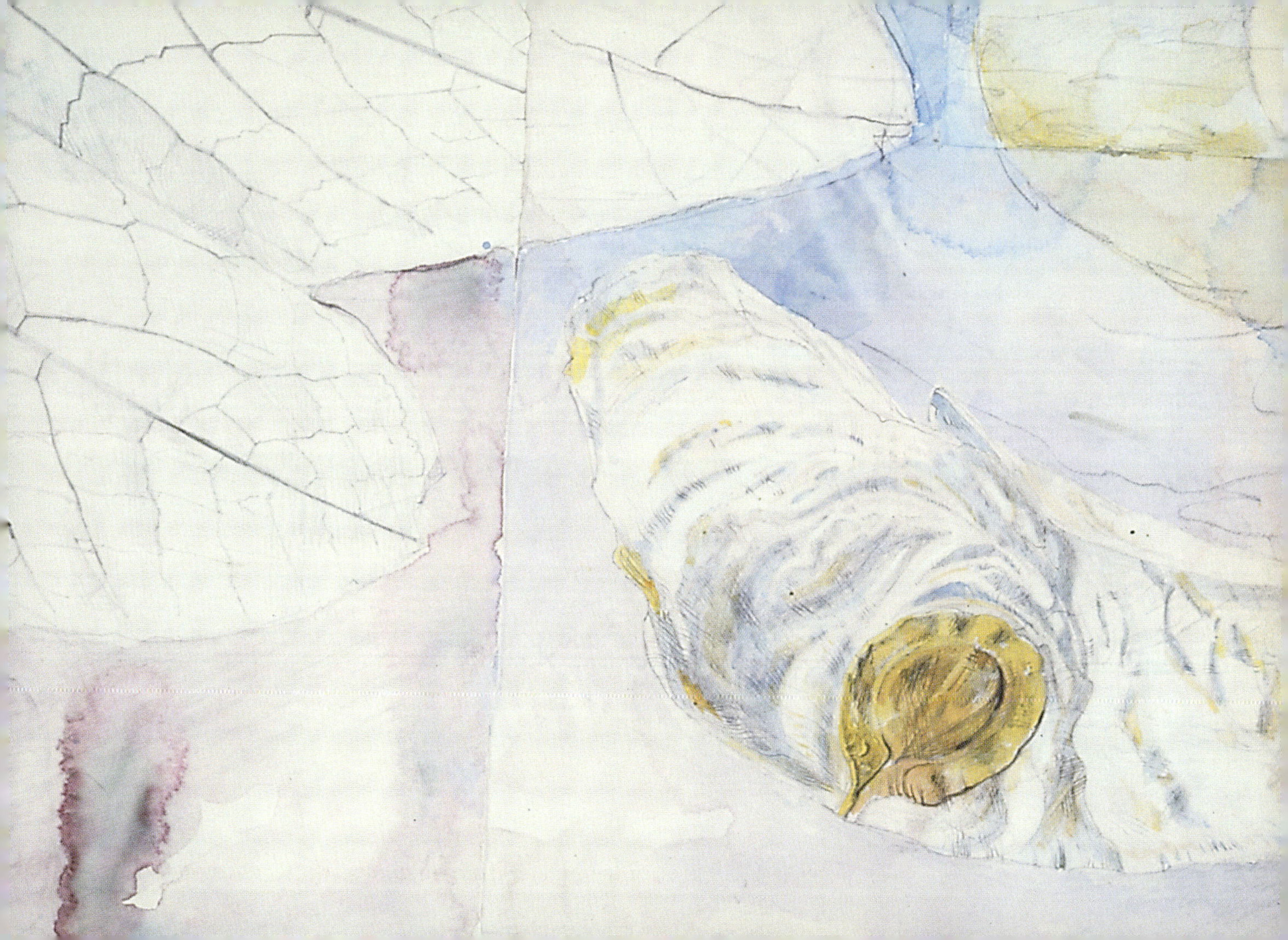

When I noticed this it seemed quite eerie - there was a complete difference between the sounds and yet some strange correspondence between them. Perhaps I was able to sense this because after dwelling so long on this mysterious axis of difference, I had some insight into the inner forces which have given rise to these different life forms.

I have now collected some 44 of these bird correspondences. Like olive-backed sunbird and scarlet honeyeater. The sunbirds, lots of them echo the honeyeaters on the Gondwana side of the Line and just this one strays across to Australia. Last year I saw them both in the same tree in Cairns both sucking honey out of the flowers of a firewheel tree. Completely unrelated species with exactly the same drinking habits.

Opposite & following page: *The architecture of leaves in Java and Malaysia* (detail & full image) 1999 (trace No. 11)

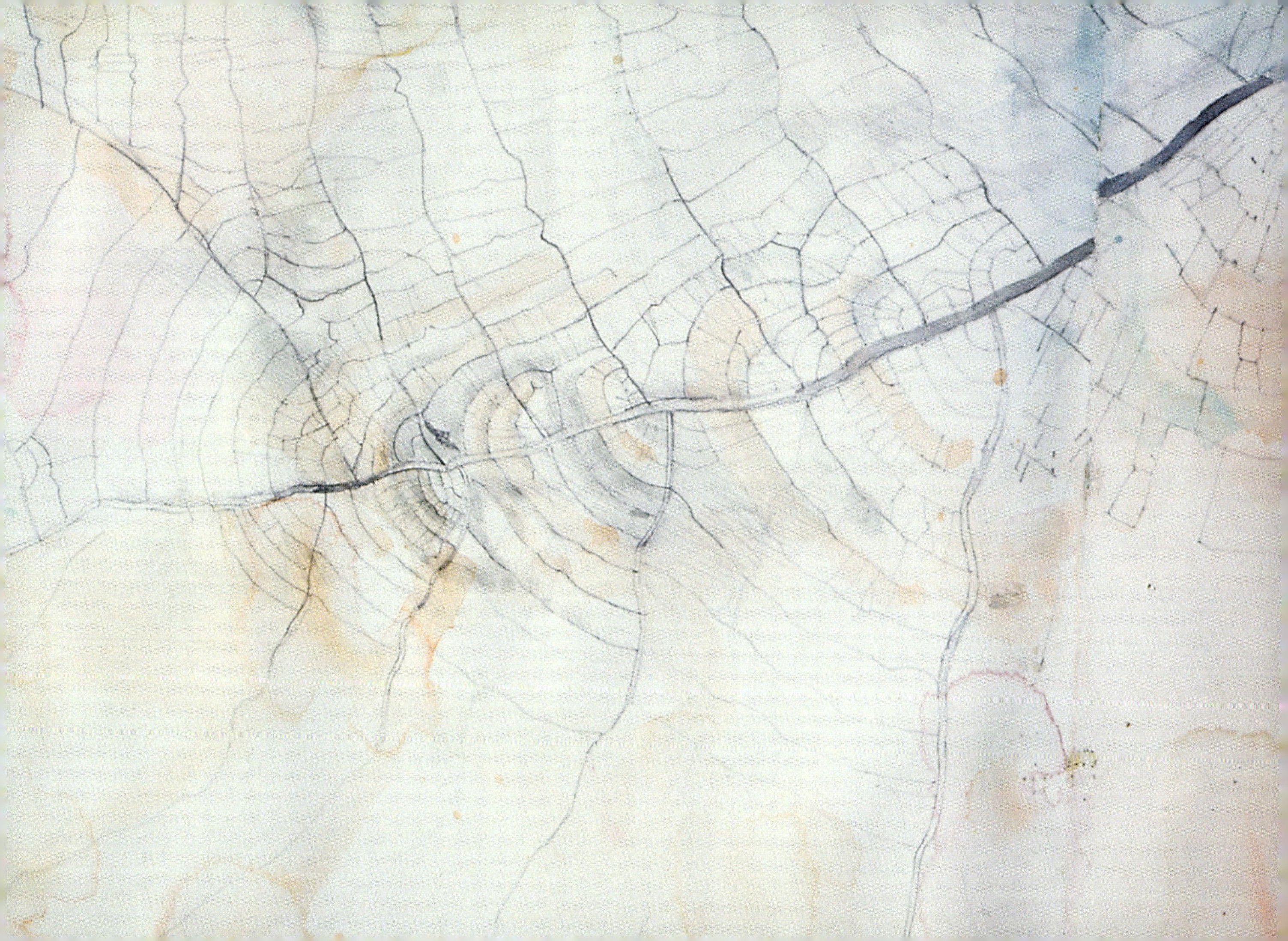

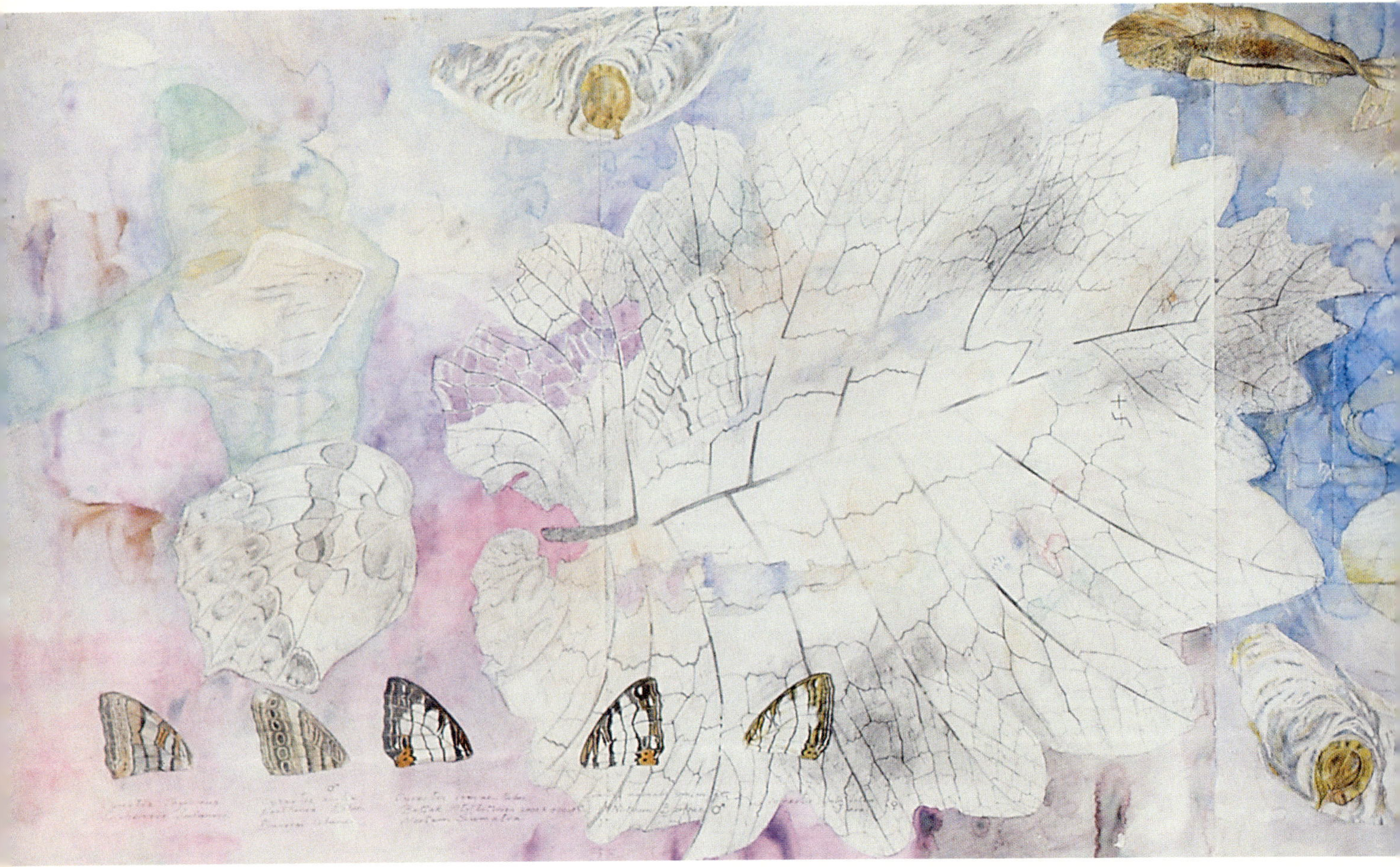

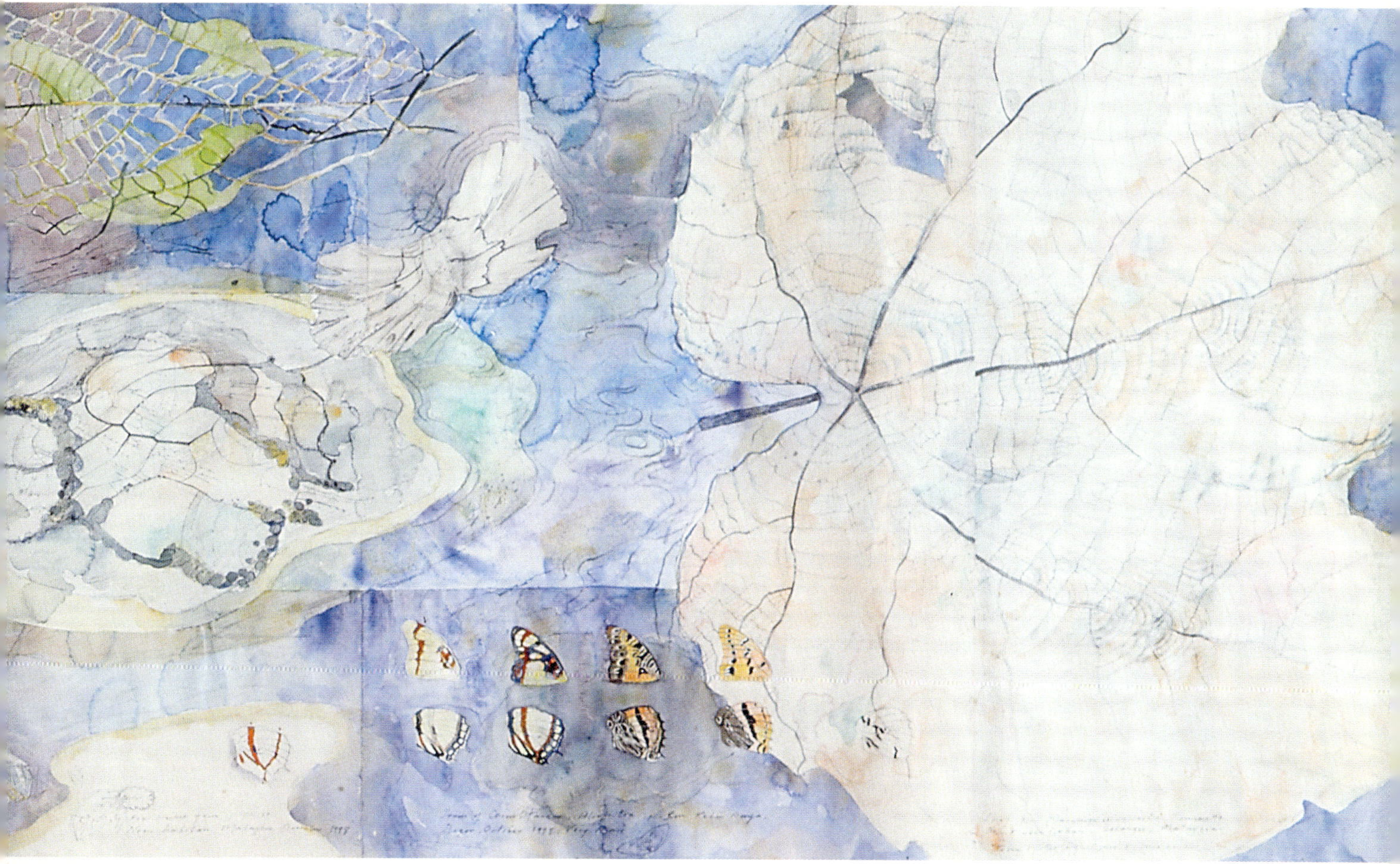

The ship & hooter & music & about to move away from this estuarine harbour with its misty islands. Now it is moving & this a beautiful place — there was a fierce-looking boy ~~[illegible]~~ who had a boat to go across. to a hamlet across the water. I think one could hire him to visit one of the little islands.

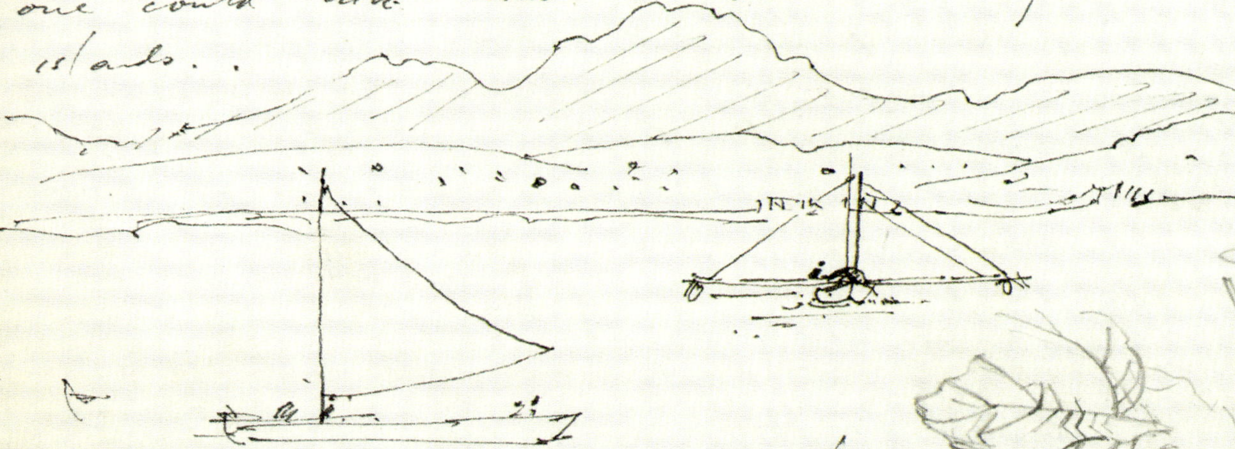

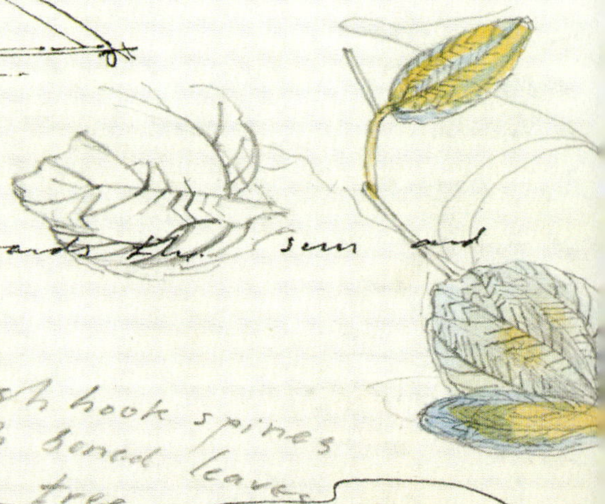

we are hugging the shore, now towards the sun and a calm ~~pay~~ pale opaline gray sea.

Fish hook spines
Fish boned leaves
Only tree on flanks
Of scorched pale hills.
Ziziphus n. Ziziphus:

We are leaving an estuarine ... & its little islands each with differences ... of flora & fauna — to arrive at one little group of Islands which has a live populations of giant Lizards still going after 130 million years — since the Jurassic period

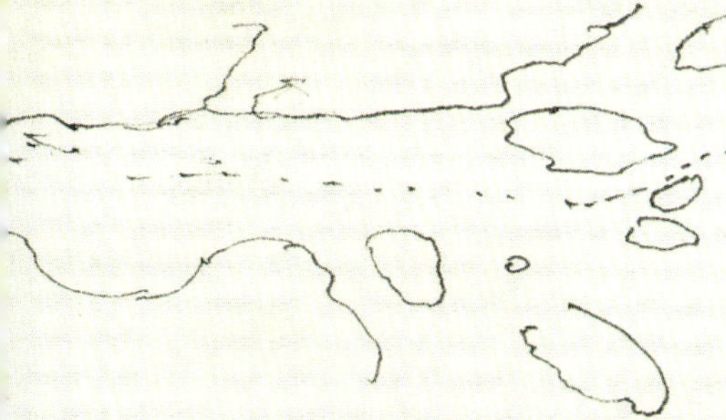

one hour out photograph 2 barren islands

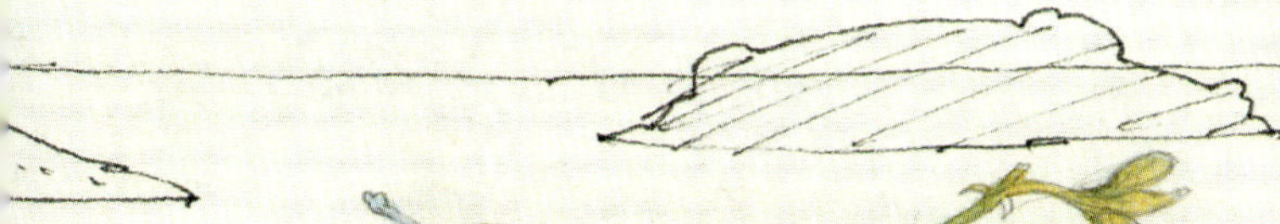

one of the

I've just looked down into the cargo-vehicle hold. there are groups down there with cardboard boxes, countless containers lying on a silky ... mattress & 3 women with a fruit market at their feet — snake fruit, Rambutan. Up here — a movie on the video about gamblers Africa ... in Black. Parallel world as the cliffs of Sumbawa float past RHS. Parallel non-... ... more and their ... world

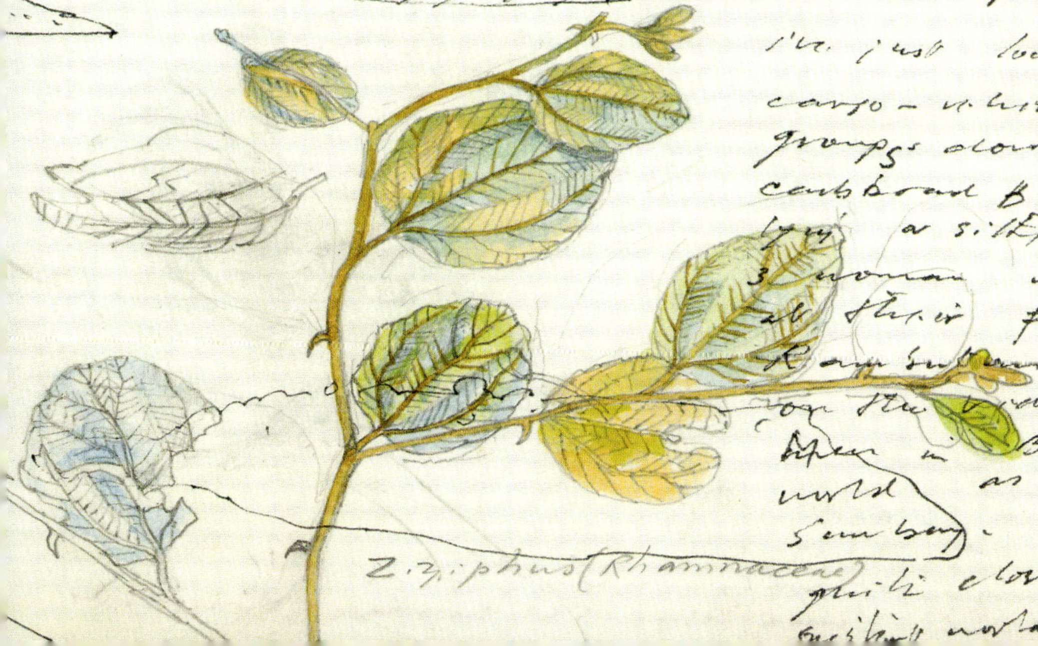

Zizyphus (Rhamnaceae)

FERRY AT SAPE SUMBAWA, 14 MARCH 1998.

Ship hooted and must be about to move away from this estuarine harbour with its misty islands. Now it's moving. It's a beautiful place - there was a fierce looking boy who had a boat - I think one could hire him to visit the little islands. We are hugging the shore; now towards the sun and a calm, opaline grey sea. We are leaving a bay and its little islands each with differences to its flora and fauna - to arrive at one little group of islands which has a relic population of giant lizards still going after 130 million years - since the Jurassic period. One hour out I've photographed two barren islands. Just looked down into the cargo vehicle hold. There are families down there with cardboard boxes, countless containers - lying or sitting on mattresses and three women with a fruit market at their feet - snake fruit, rambutan.

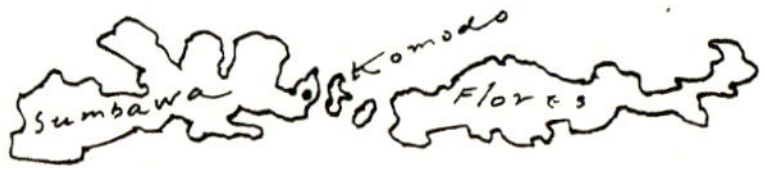

Opposite: *Macaranga drift* (detail) 2001 (trace No. 14)

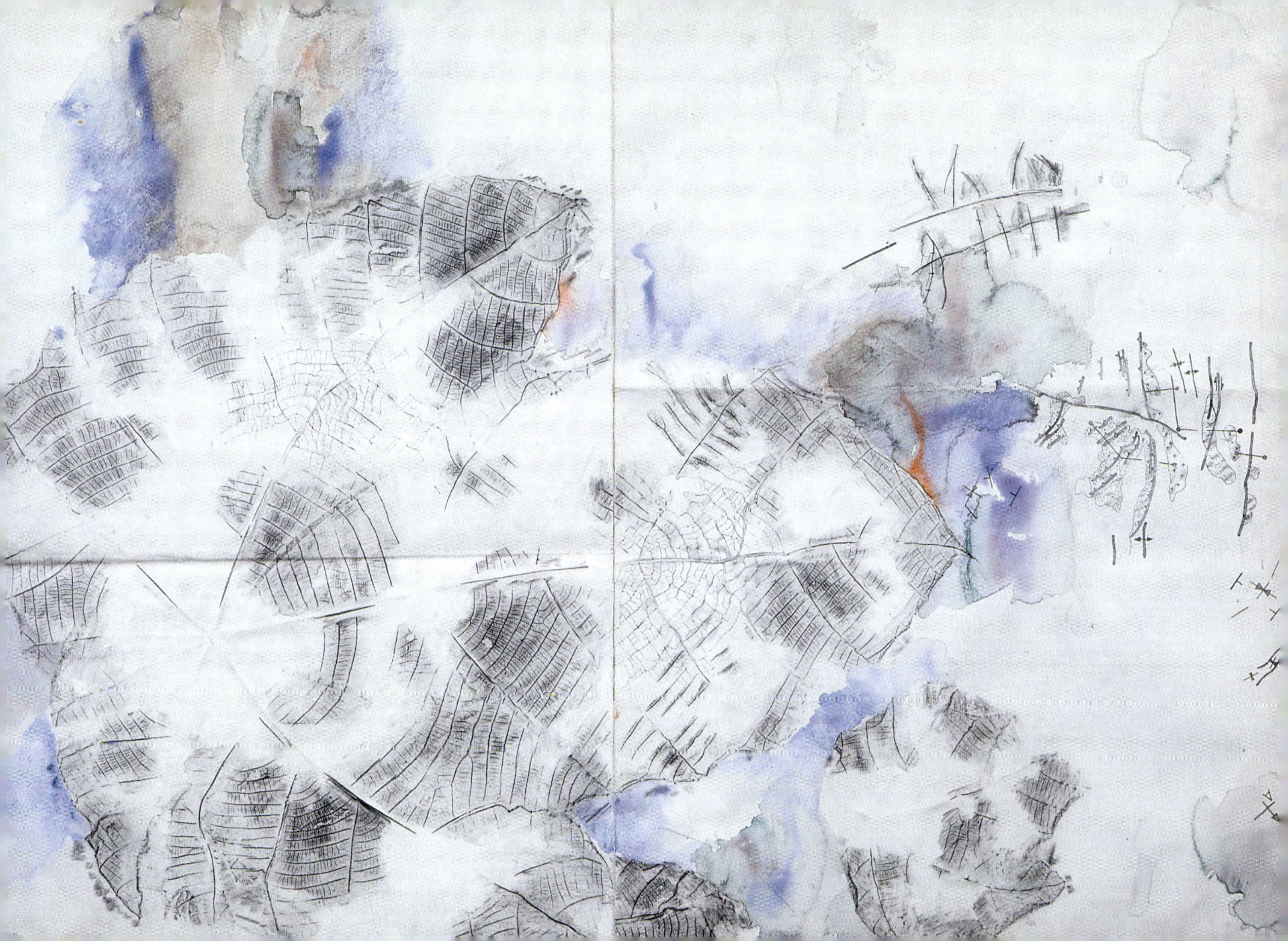

Up here a movie on the video about Aliens. *Men in Black* - a parallel world to the cliffs of Sumbawa that we move by quite close. A parallel non-existent world more real than the one on board the boat - a world passing away. I think we must be reaching the end of Sumbawa - cliffs I photographed, then big barren lolloping island. Very rich tawny with look-out (military?) on top. And dramatic synclines and folding rocks like teeth coming out of the sea. Took several photos with polarising lens - images of the colliding margins of eroded continental plates and bits of plates. Ridges of the earth's crust, overturning and collapsing. The aliens are tumbling into the backs of garbage trucks. People firing guns at them and falling back into other garbage trucks. "he's dead... he's dead!" - Bursts of gunfire above the growl of this rust bucket's engine. Wonder what all these images and chaos and smashing things up means. Perhaps indicative of a deep down fear that we have lost our place in the universe by destroying it, so we construct this violent screen as a displacement activity.

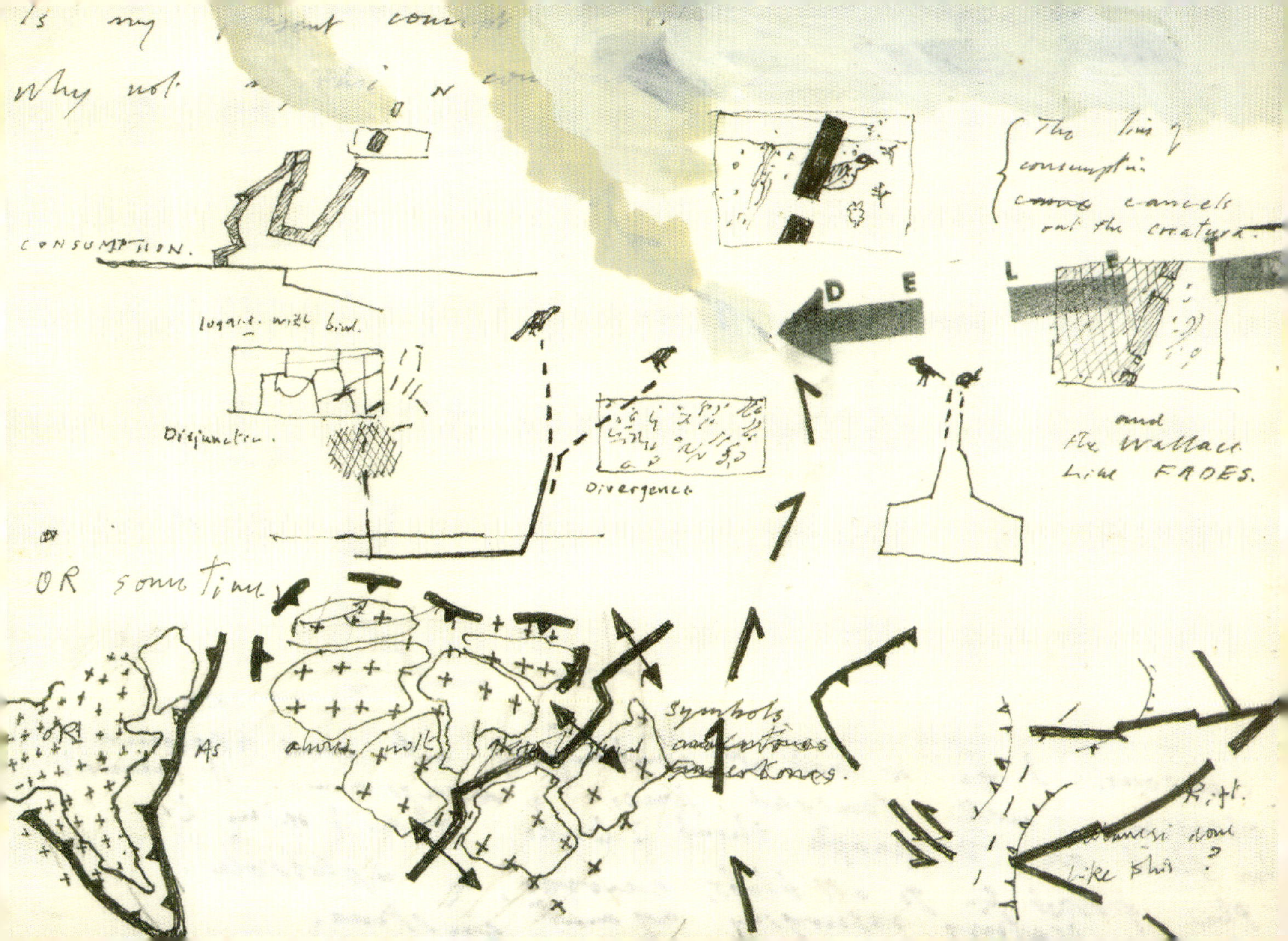

why not
CONSUMPTION.
Disjunction.
Divergence
The line of
consumption
cancels
out the creature.
DELETE
and
the Wallace
Line FADES.
OR sometimes
As above
Symbols
Rift.
like this ?

We sort of fell or spilled out of the small fishing boat which had taken us to the Komodo jetty. Several dragon voyeurs and tourists. There at 2pm the hot savannah and jungly trees and bungalows on stilts looked down on a beach with a lone Ruisa deer with its elegant legs resting in the cool margin of a calm sea. Got key to my room - only me in the wooden stilt bungalow - me and many geckos.

My shirt is permanently wet from the heat. Bird noises, no wind but a flutter from tree to tree. Green-winged pigeon cooing. Friar bird "eeyow-oweeayowoyo." Dry dusty forest but with some jungly trees with big leaves drooping in the heat like flannels hung up to dry. An ashy drongo was sitting in a tamarind tree. Weird song - musical and yet "zit...zit" not. A helmeted friar bird - knobbly on a knobbly twig - called to another "eeyour-oweeayowopup."

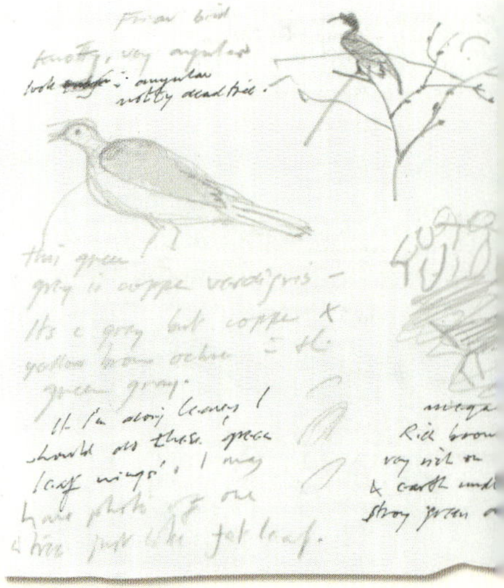

Splosh cold water from Mandi bucket - ah - that's better - new shirt on again - soaked within minutes. Open air canteen - onion omelette. I've taken *Birds of Wallacea* and about six park guides are poring over it as I eat my omelette. Their discussion is like some weird bird language because they know all the scientific names so I can hear scraps of Latin mixed with Lehabangan exclamations and strange pronunciation of common names "Ah... whit napped uril" (oriole).

Maen, Main bird expert ranger wakes me at 5.30 armed with panga knife and large fork for defence against dragons. We pound quietly up the dry earth track. He stops and gestures at two white imperial pigeons still seeming to be asleep high in some branches, and then towards the huge dome of a ficus tree. Behind it the sun is coming out of some parched reptilian hills curiously studded with Lontar palms.

I look at the tree with my binoculars and they reveal a break in the round canopy from which glows another sun - two arcs of the purest yellow - cadmium yellow wings edged with black held open to receive the warmth of the sun now rising into the sky. A black-naped oriole on the rim of this dark mass preens and shakes the black stripes across its head and tail, and shivers its yellow wings.

I did some drawings of the bird whilst Maen stood near me - forked weapon at the ready. "There's lizards round here" he said. I couldn't hear what he could hear until there was a crashing of undergrowth and clouds of dust. Something like a berserk truck crashed over the side of the ravine and hurtled down it followed by a something as big. I caught a glimpse of black mangy hide and then a large flank of scaly muscle. "Dragon after buffalo" Maen said. We then went down to a wooded valley with same dead imploded palm trees like collapsed archaeopteryxes and big feathery cycads.

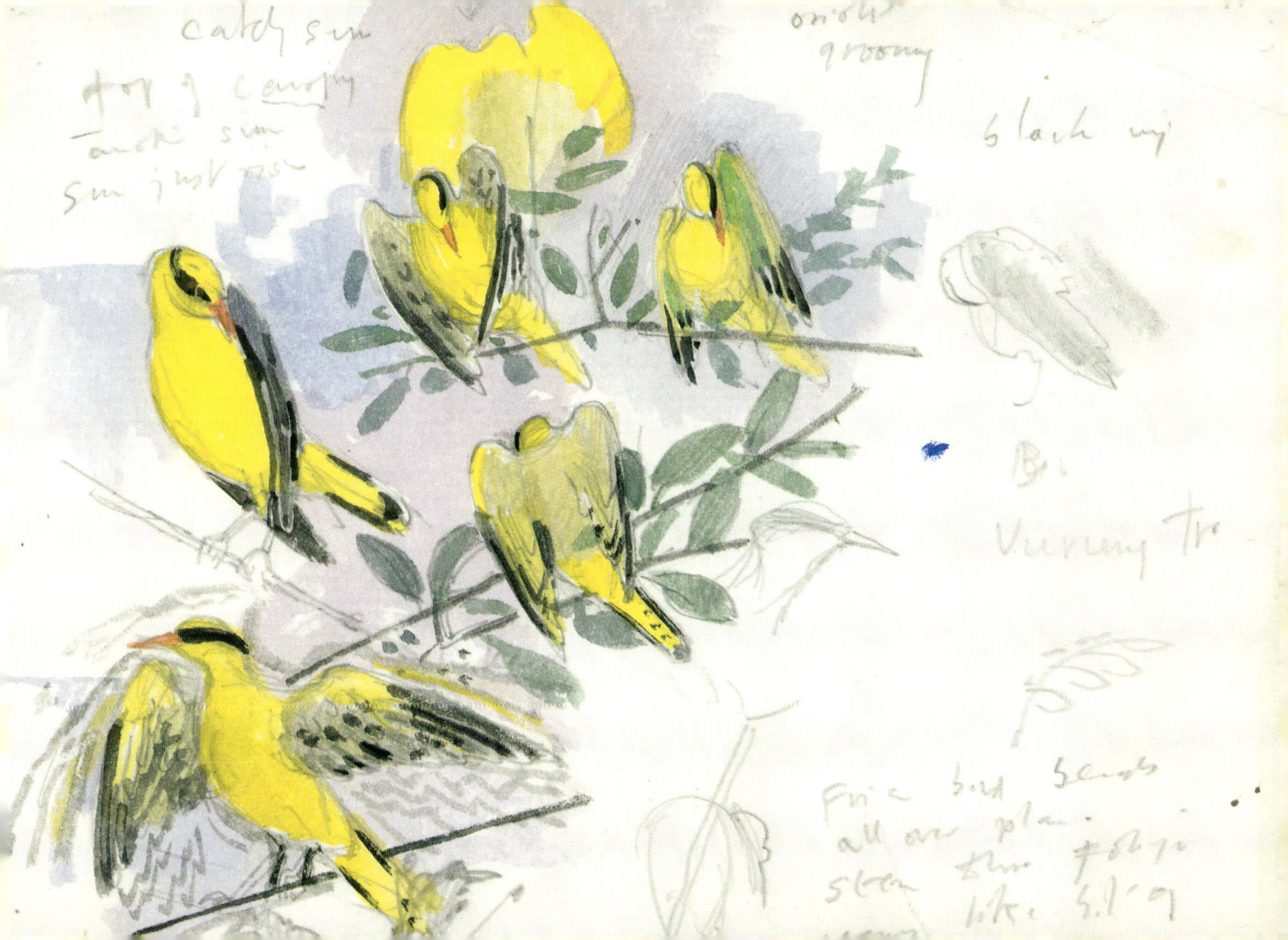
catch sun
top of canopy
oriole
grooming
black wing

In the branches of a ziziphus tree there were several drongos by a nest - such fishtailed elegant shapes - so mercurial and such peculiar songs. At the same time several pied imperial pigeons alighted in a kapok tree (I think) and I got into a totally apoplectic state - drawing, swapping different lenses - knocking over the telephoto on its tripod - stalking the birds - panting with heat.

Now I'm back in my stilt hut with a foot long gecko, its huge eyes watching me from a rafter - an olive and brown pattern on its back strangely similar to the feel of a Sumba ikat weave. Against the pale green twilight sky, there is the friar bird still making its strange cries - as if continuing the same story sung by that other friar bird four months ago by the Tiwi sea in the far North of Australia. Down below, a Komodo dragon has just emerged out of the Jurassic forest. As it walks slowly across the clearing its feet cause little puffs of dust illuminated by the moon.

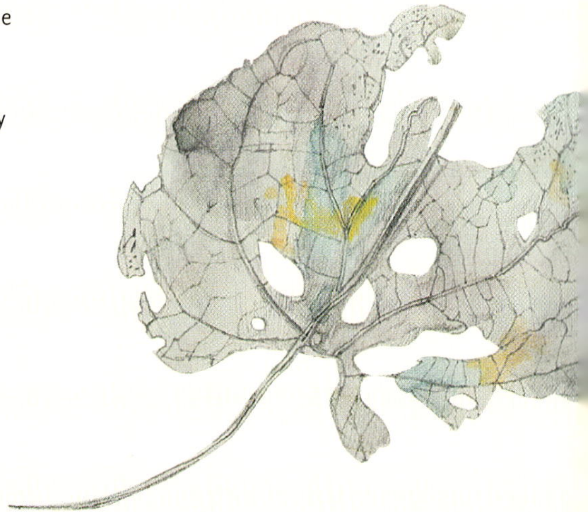

This afternoon I was walking along, all my senses awake to this strange habitat and the unfolding of time - and then I passed through some invisible skin or boundary into a quite different time world. It was like taking off the papery layers surrounding a nut, and finding different coverings, or even unwrapping the rock stratas of the earth into deep time. There are places where there are discontinuities between where one is, and where a layer has been folded back to reveal an unknown or archaic zone. And perhaps not only in one's surroundings - crouching as I was under relic cycads - I was thinking this could be true of one's mind, one's psyche. It may be because I am so sorely missing my bread and croissants - after two months of rice in Indonesia - that I've been thinking perhaps my psyche is formed in something of the manner a bread maker rolls and folds the dough, turns it round and kneads and folds it in on itself again and again and so on.

As with pastry there is no steady accretion of the same stuff, just a marvellous interleaving and enfolding of different parts - systems folded into different systems and dimensions (currants?). There are unknown elements and worlds enfolding within the dough - intimations and ancient memories which break through all the preconceived half baked ideas and suddenly, shatteringly open up to shifts in understanding as explosive as that giant lizard's descent into the gully this afternoon.

Opposite: *Javan leaf and Tiwi leaf , continental drift* (detail) 2001 (trace No. 1)

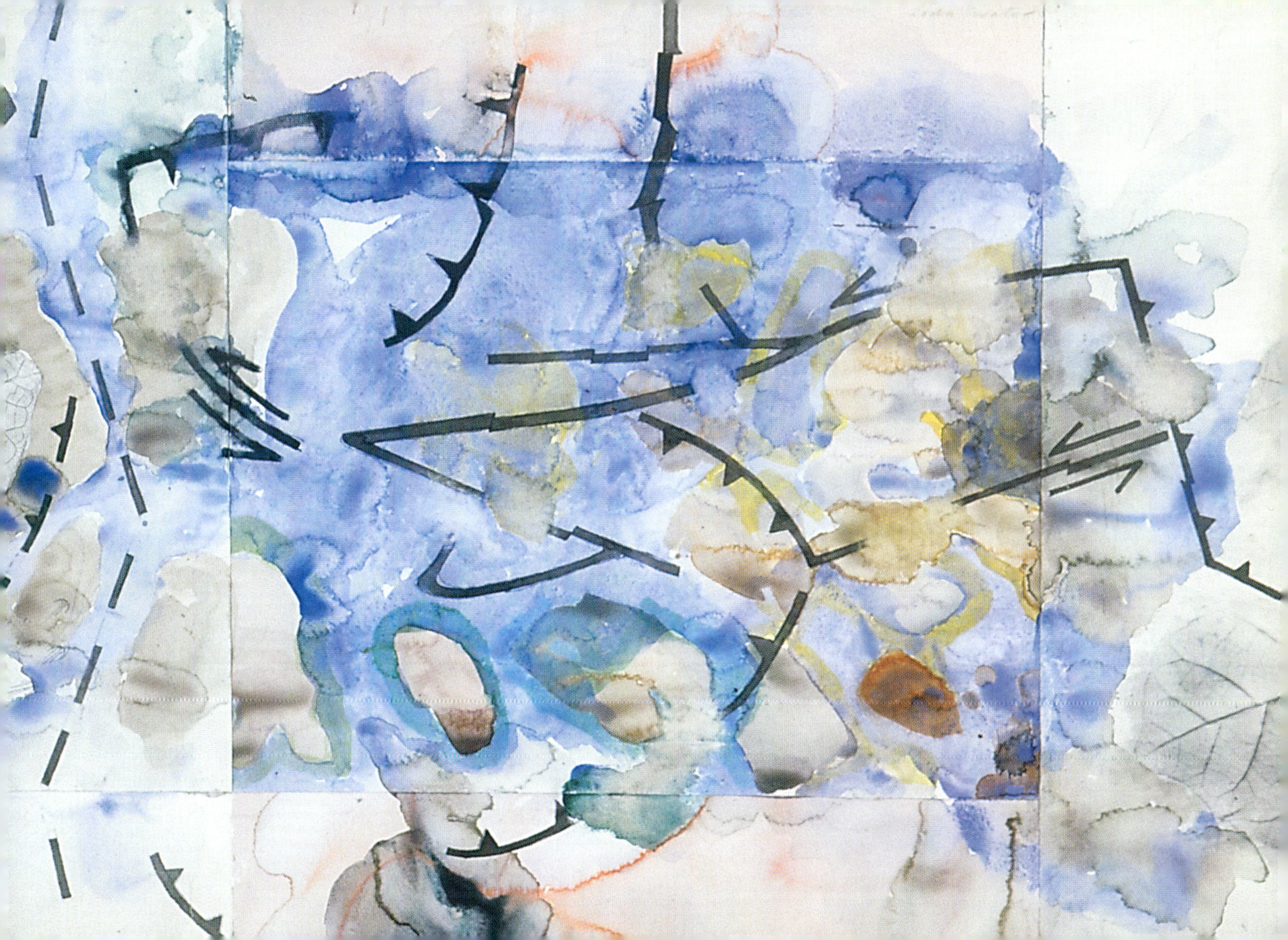

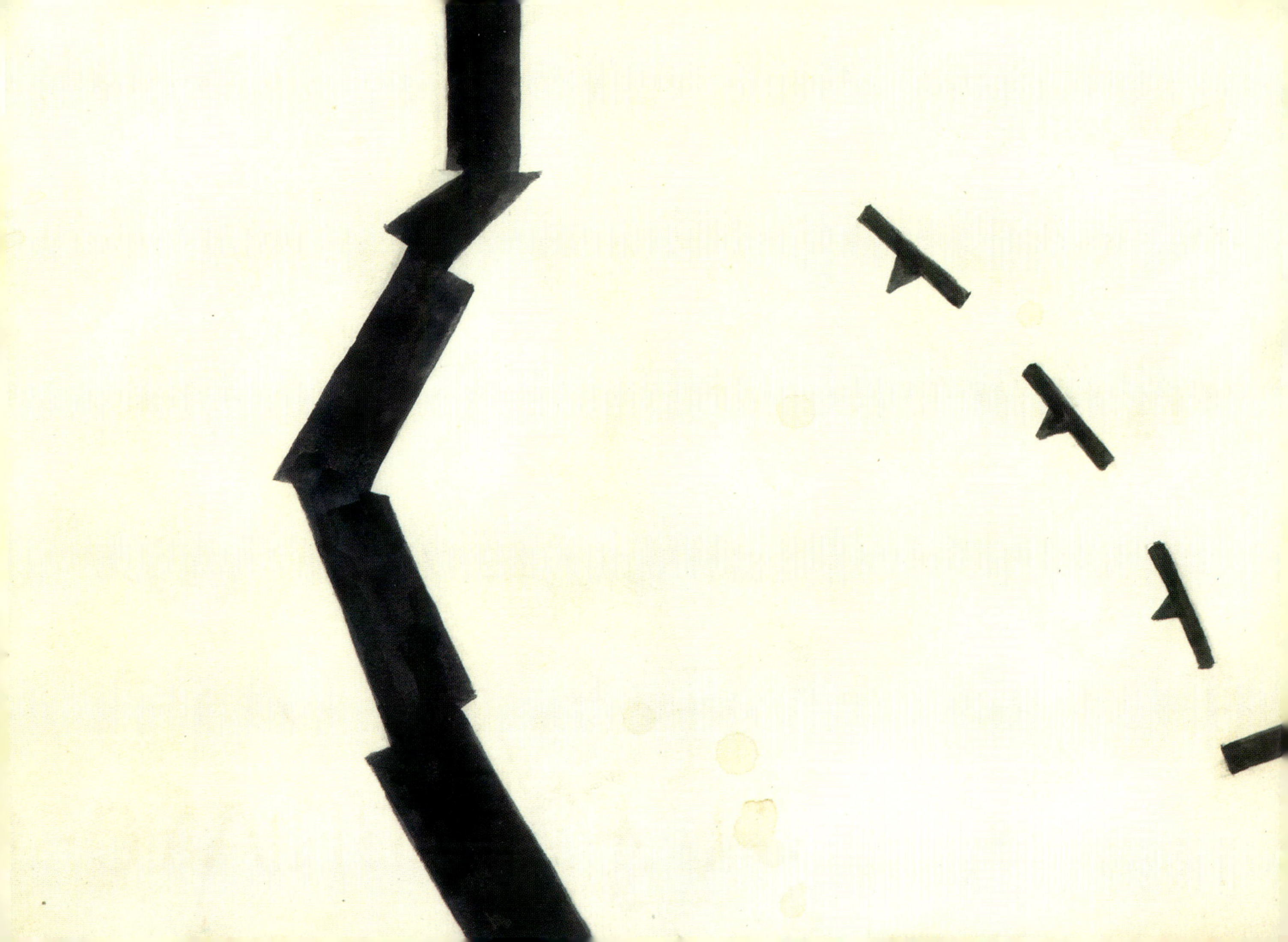

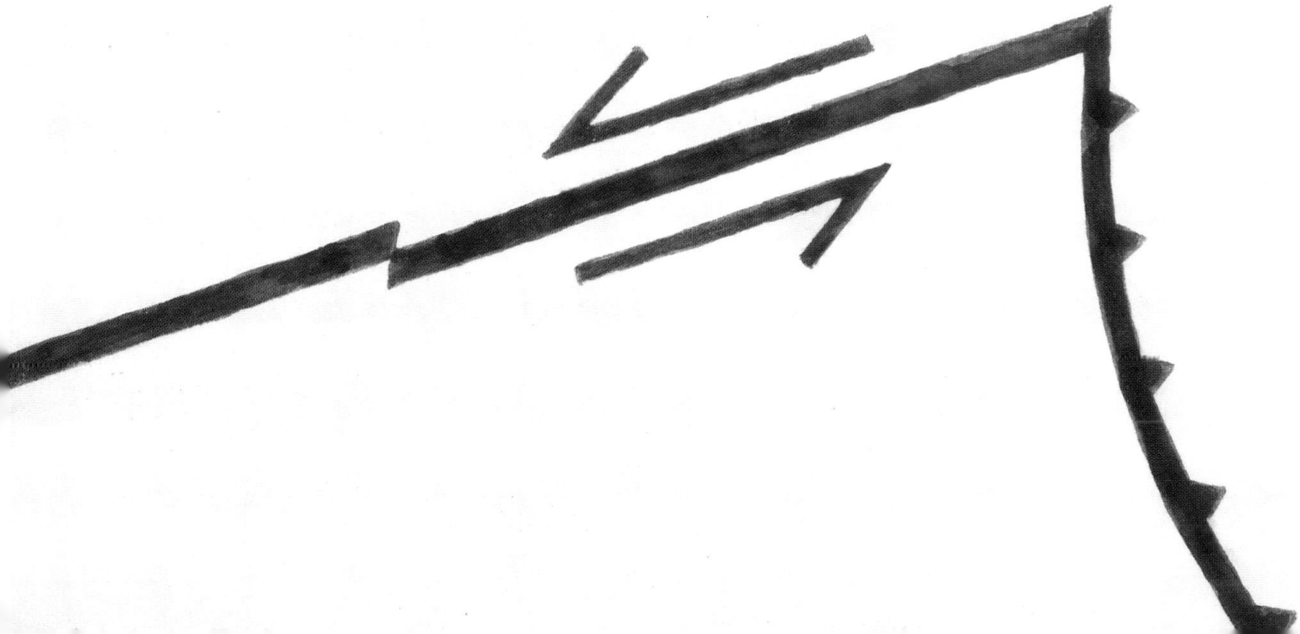

CROSSED THE LINE - ABU'S PENSIONE, BOGOR, 12 FEBRUARY 1998.

It occurs to me as I do rubbings of these leaves that they could be incorporated into my mapping drawings of land - they look like the surface of the Earth which in a sense they are. The Earth and its rock stratas are wrapped in a green leaf. And this green covering exhales a thin envelope or bubble of air. If these filigree pale images were laid on the islands of this archipelago they might seem to have a tender affiliation with each one. They might seem to be the surface skin. Intimately connected. The painting might hint that - say in Java - the dipterocarp layer has mostly gone. It is as if its natural skin has been torn off.

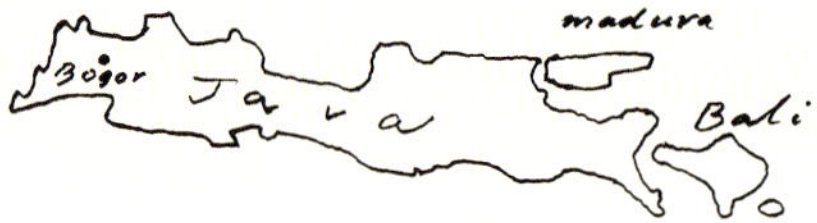

It occurs to me as I do rubbings of these leaves that they could be incorporated into mappy drawings of land – they look like the surface of the earth which in a sense they are. *

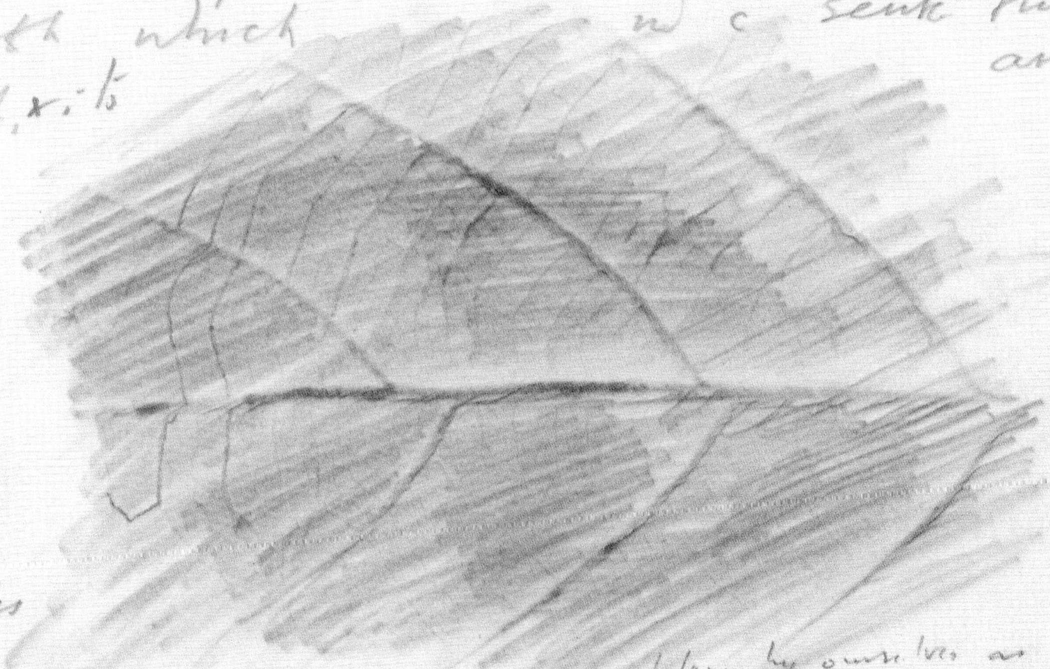

The earth, x it's the rock stratas are wrapped in a green leaf. And this green covering exhales this envelope or bubble of air. blown by ourselves as Cf Lewis Thomas
— 'We live inside a blue chamber, a bubble of

Spent last few days drawing plant specimens in the Herbarium. All these cabinets packed with pressed plants from the far flung islands of the archipelago. Yesterday I was drawing a large winged dipterocarp tree seed. Dr Harry Wiriadinata, the Curator, looked over my shoulders and said I should treat it with care. "It's the last known specimen from a tree which is probably extinct", he said . The delicately faded fruit with its long slender wings then became in my hands an object of such poignancy. I wondered if there will come a time when most of what remains of the natural world will be found in the cabinets of museums. I must discuss with Linda Fredheim the idea of making some cabinets which are in memory of lost species. I think the top section of one of them might be a kind of casket which has the feel of a reliquary case or sacred container.

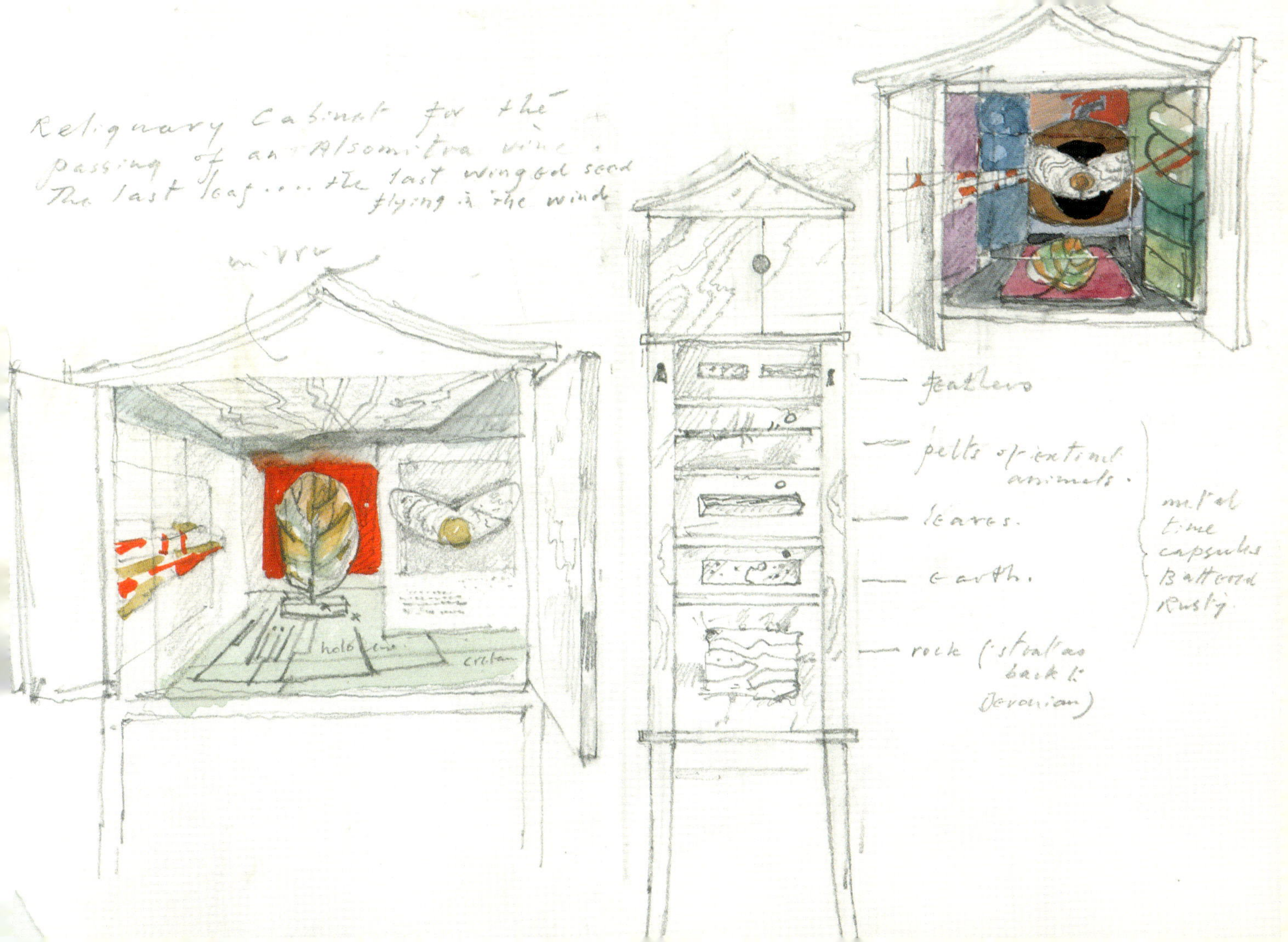
Reliquary Cabinet for the
passing of an Alsomitra vine.
The last leaf ... the last winged seed
flying in the wind
mirror
feathers
pelts of extinct
animals.
leaves.
earth.
rock (strata as
back to
Devonian)
metal
time
capsules
Battered
Rusty

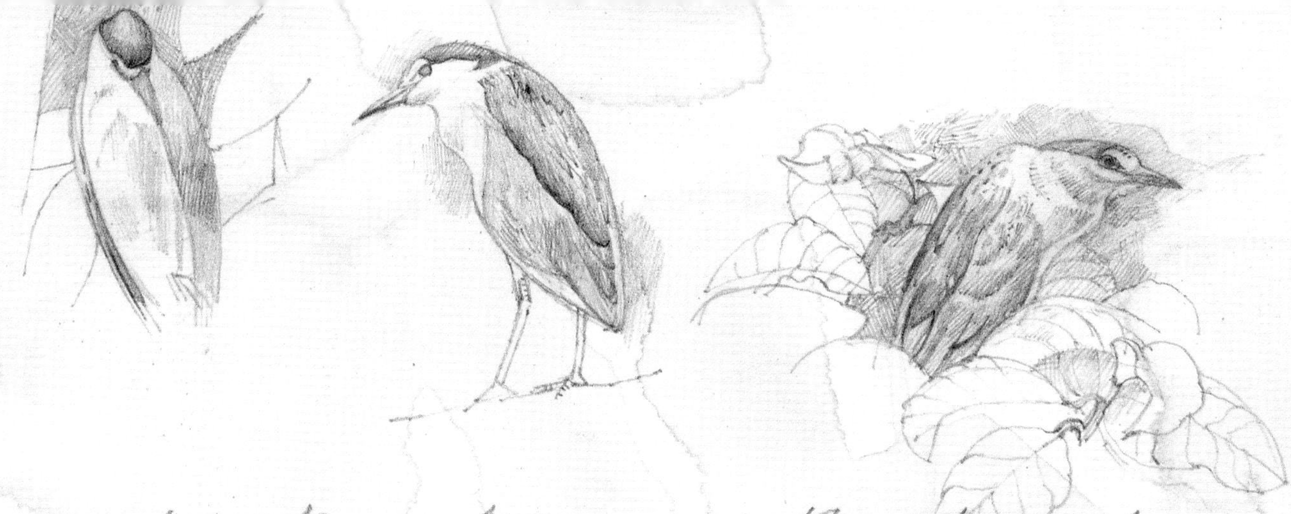

moonlit water. I imagined they had always
roosted on that natural lake there, in that
sanctuary – long before ~~[illegible]~~ this city sprawled around
it. And for eons they had each dusk flown South
and East to their feeding grounds. ... still moved
in secret accordance with the spread & drift
of the earths' crust – following the memory of
the ~~[illegible]~~ patterns & force fields which had
evolved long ago when the Laurasian continent
was so much further North

I moved to a room which overlooks the ravine - sound of river and not so much clatter. Then walked to Botanical Gardens through the packed streets. Somewhat ominous groups near the civic buildings some with head bands with slogans. A sentinel night heron on a branch guards the islet on the lake. Others I could see through the telephoto roosting within the trees. The dark recesses within held these beautiful long oval shapes. Twilight came. It was surprising how many burst out of the interior making strange croaks.

They circled up above the Keban Raya - and this vast town - and flew beyond the hills. I thought of them standing motionless beside distant paddy fields and lakes - watching for fish in the moonlit water. I imagined they had always roosted on that natural lake there in the sanctuary - long before this city sprawled around it. And for eons they had each dusk flown south and east to their feeding grounds.

They still moved in secret accordance with the spread and drift of the Earth's crust - following the memory of the patterns and force fields which had evolved (long ago) when the Laurasian continent was so much further north.

I went down the path through the huge clumps of bamboo past the cemetery for dead birders (Kuhl etc). A cat lay on Kostermann the botanist's grave. It yawned and watched me pass as I walked to the south entrance and had a nasi goreng along Surya Kancane.

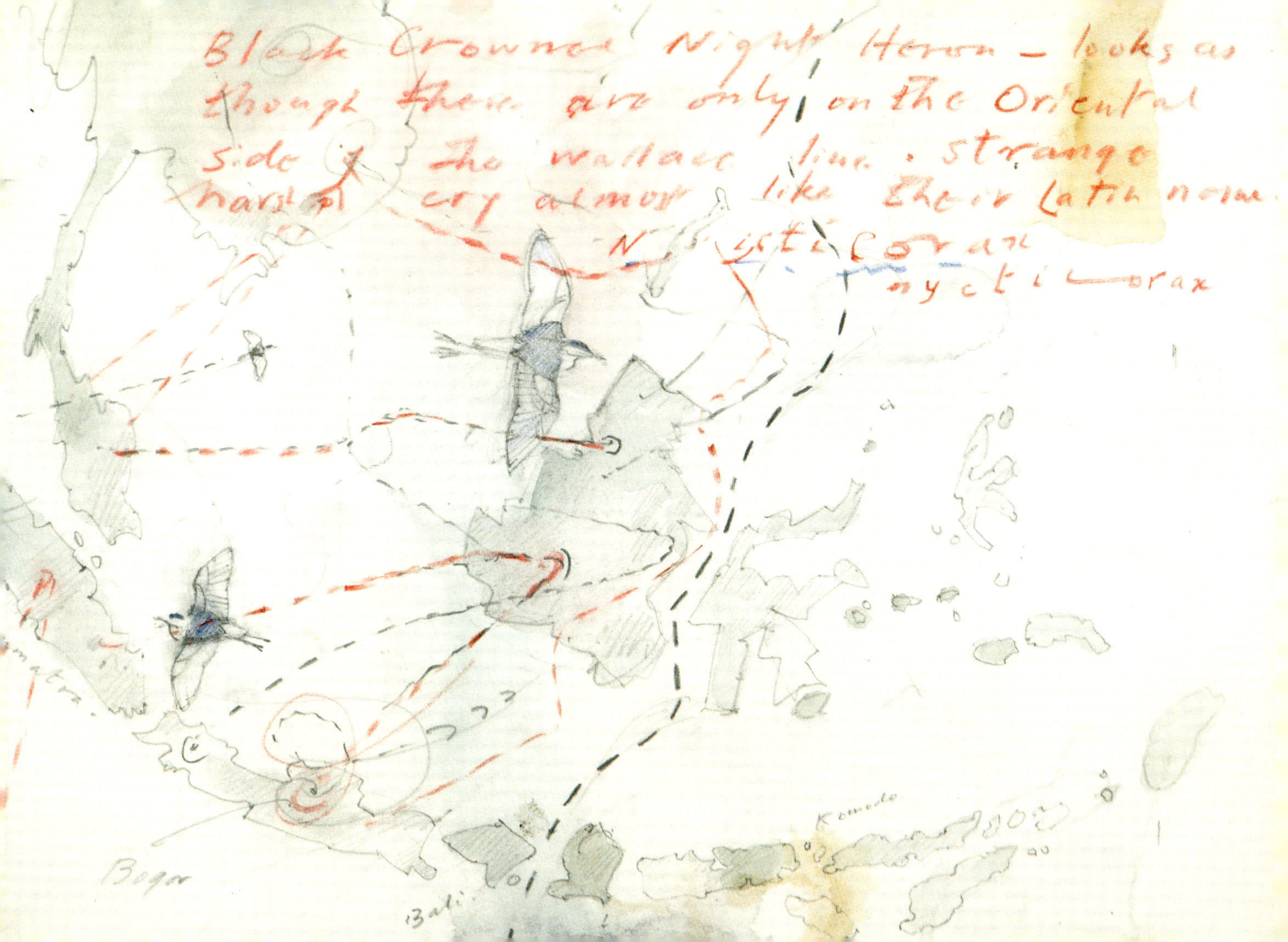
Black Crowned Night Heron – looks as though there are only on the Oriental side of the wallace line. strange harsh cry almost like their Latin name.
Nycticorax
nycticorax
Komodo
Bogor
Bali

KUALA LIPIS, 7 DECEMBER 2000.

A small town, a kind of suburb of Kuala Lipis. There's a relic bit of old jungle with a hide described in the tourist information. So last night I went through allotments and some re-growth by the river and entered some deep jungly stuff. From the shallow veneer of human settlement it only takes a minute to be enveloped in that total world of dark trees. The energy of this tropical vegetation even when only allowed a small area is still astounding. As in the Green Mountains and Cyclops Mountains I felt again how powerfully completely 'other' this forest dimension is. It's not so much alien, as in the dark world of Conrad, but has its own state of being to which we have not been given the secret. I find myself falling just short of some strange belonging. I followed a track through buttresses of trees and several microclimates of warm or cold or scented air and reached the wooden stilted hide overlooking a partly mown clearing.

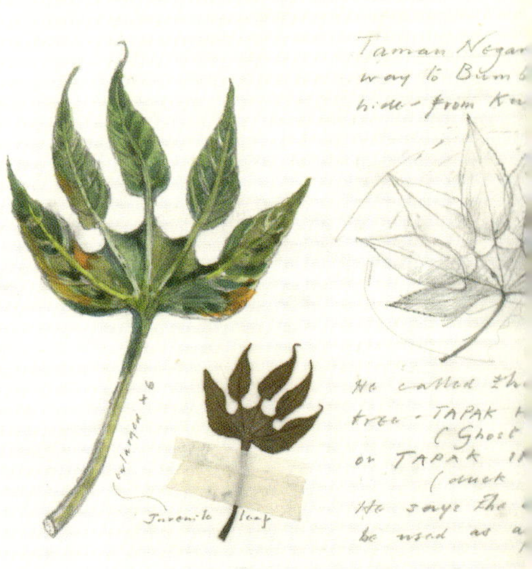

The tourist notice said that wild boar, tapir, monkeys and an occasional tiger could be seen from here. A very occasional tiger I thought. It was mysterious peering through the long slit in the side of the cabin at this patch of ground surrounded by strange jungle trees with a forested hill behind. As it became darker with my mind prepared for strange manifestations, I was almost seeing imaginary animals as I gazed at where the open ground met the dark scrub. One or two of the huge jungle trees themselves exuded a presence of something animal and intense. A boy of about fifteen climbed up the steps and sat down on the bench. I said "Selamat petan" and he answered "Good evening Mister, where you from?" We sat down on our bench silently staring. He beckoned at the ground below where three squirrels (shrew-nosed squirrels?) were chasing each other - short fast runs up tree trunk and down again. The moon came up over a dense great mushroom shaped tree and we watched the illuminated space below.

After half an hour I asked him if there might be tapir or deer ? "No... nada", he said "Not any wild boar or monkeys?" "Long time gone" he said.. "Perhaps the occasional monkey?" "No...only squirrel" he said.

It was sad but also quite funny, like some kind of Pinter play. Here I was in a wild life hide - on the edge of the wild - looking at the place where the wilderness wasn't. The building was for looking out from - with its long horizontal slit - looking out and being hidden from the wild creatures. In a strange way because there was nothing out there - no wild things - I felt a strange sense of exposure and vulnerability - a kind of loneliness.

Opposite: *Tracing the Wallace Line: wing leaf and land* (detail) 1999 (trace No. 6)

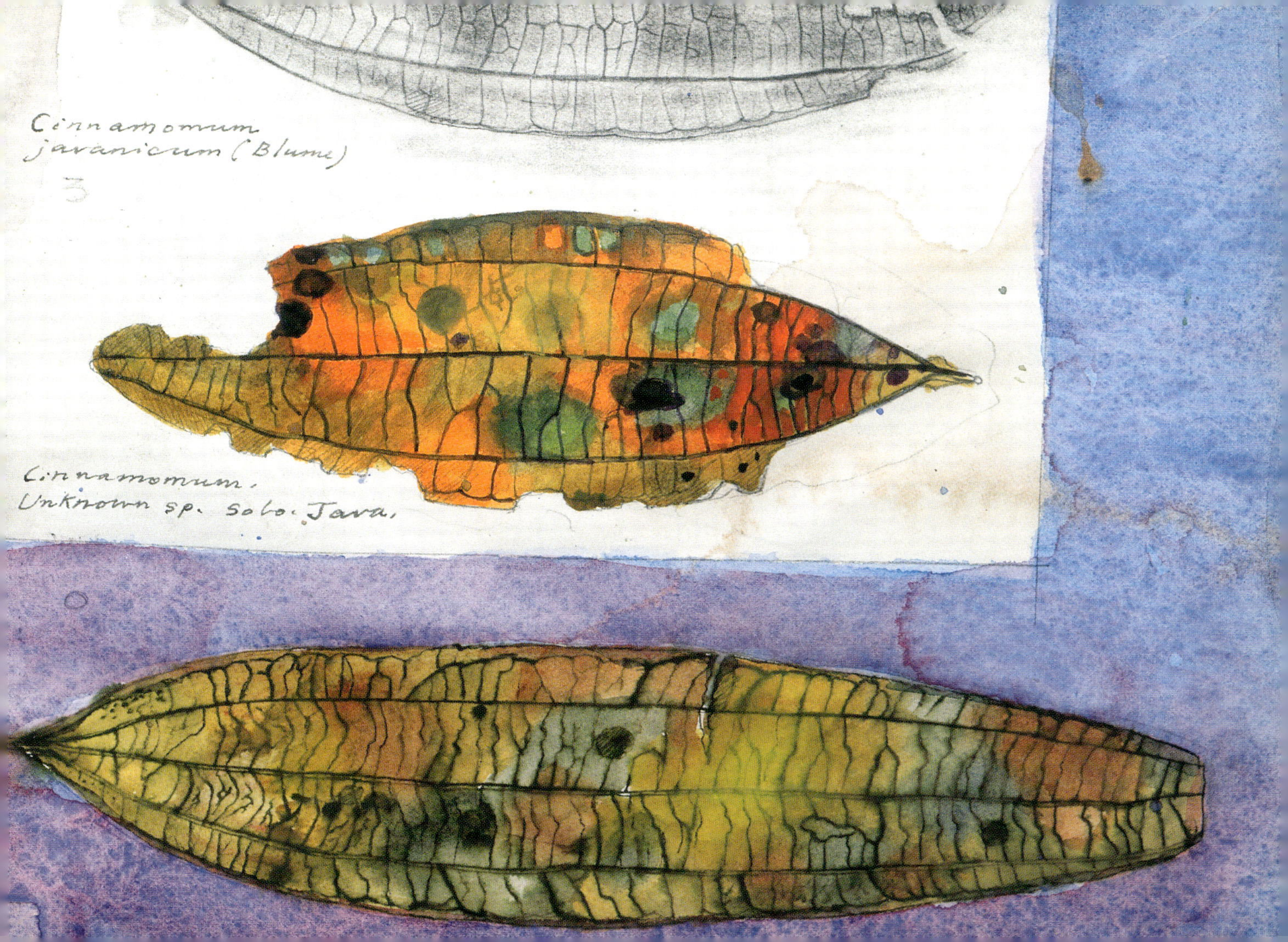
Cinnamomum
javanicum (Blume)
3
Cinnamomum.
Unknown sp. Solo. Java.

TREE CABIN, MESILAU, MT KINABALU, 3 JANUARY 2001.

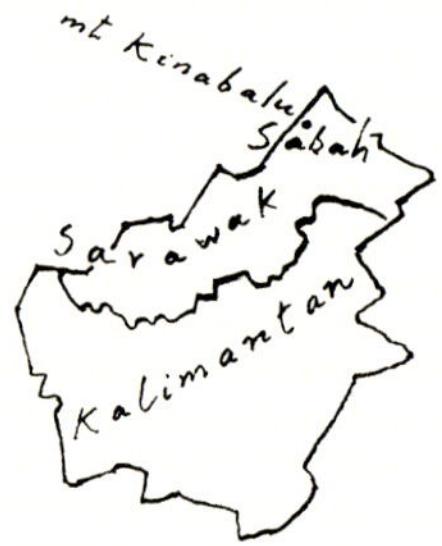

Lying in bed with shocking flu - my room resting in the branches of the ancestor oak-(fagaceae) trees looking at the intervals between the limbs and the filigree twigs - like the veins of the sky…I was thinking how much this project seems to be about the branching of things - tracery of trees and the pattern of leaves. Linear forms which unfold and create different frameworks for dwelling.

I lie in a kind of architecture within which various life forms live and move and have their being. From the smallest cell or the double helix of DNA with its two chains wound around one another to the giant twisting branches of a lithocarpus, enclosing monkeys, squirrels and trogons, there is a structure, a framework of an extraordinarily refined form, - one perfectly adapted to its purpose.

Opposite: *A history of ferns with hoopoe and hooded pitohui* (detail) 2000 (trace No. 12)

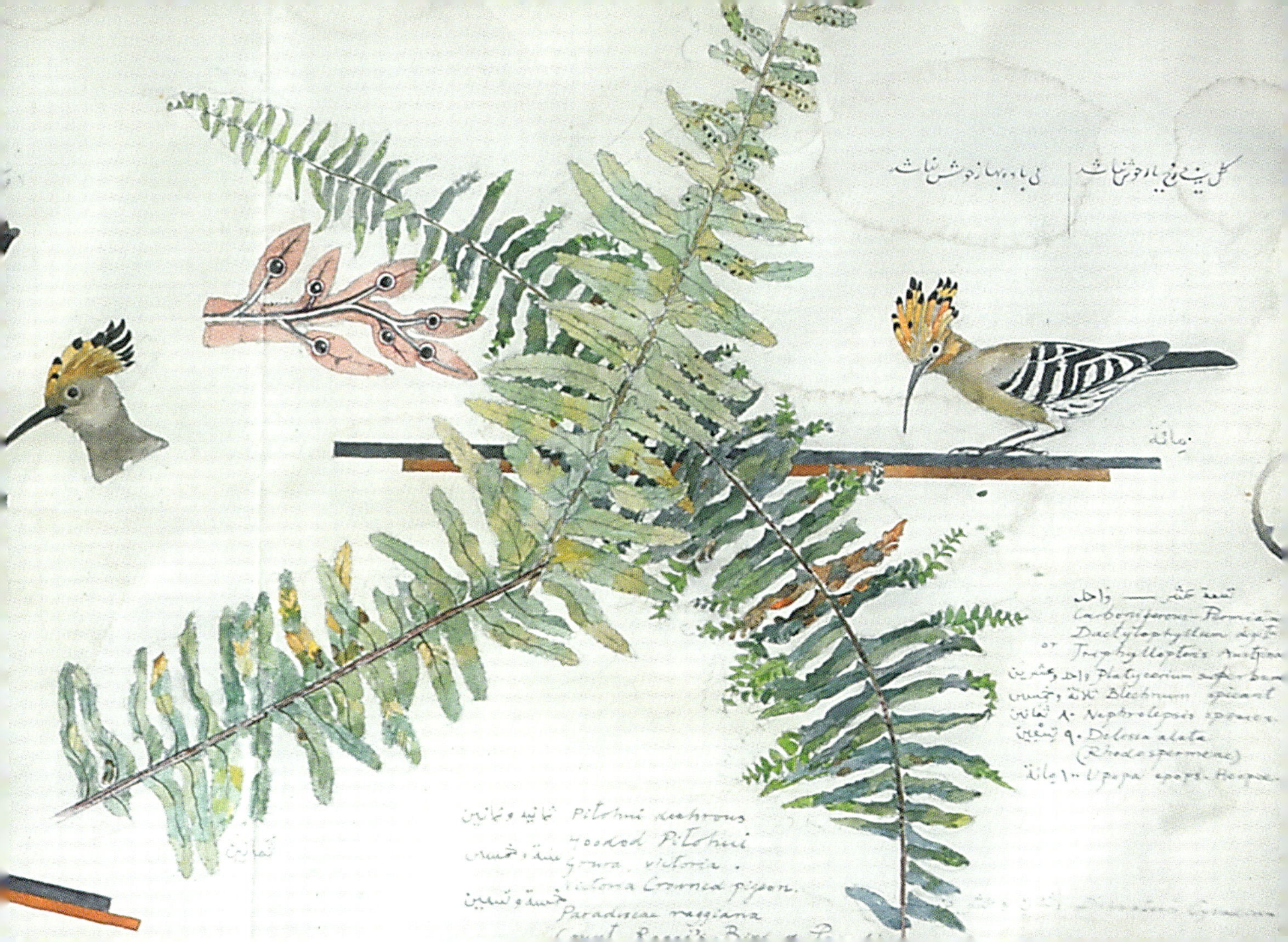

Carboniferous–Permian
Dactylophyllum
Platycerium
Blechnum spicant
Nephrolepsis
9. Delessa alata
(Rhodospermeae)
10 Upupa epops. Hoopoe.
Pitohui dichrous
Hooded Pitohui
Goura victoria.
Victoria Crowned pigeon.
Paradisaea raggiana

I think a human figure moving within a well-designed building (let's say one of those Malay Kampung houses), has a similar kind of beautiful relationship with the larger structures around it - sky, wind dynamics, geology.

As I've wandered between the islands of this archipelago, I seem to have gained a heightened awareness of the differences and similarities between the diverse tree genera as they 'express' their particular habitats. I cannot help noting how for instance, the branching of myrtle beech trees in Tasmania's mountains has some of the rhythms of oak/beech ancestors here at Mt Kinabalu. And I'm building up a kind of thesaurus of basic branch patterns which owe their form to such varying factors as climate, temperature, wind, aspect etc. Moving from the dense jungle to the more pared down mountain tree forms up here invites these comparisons.

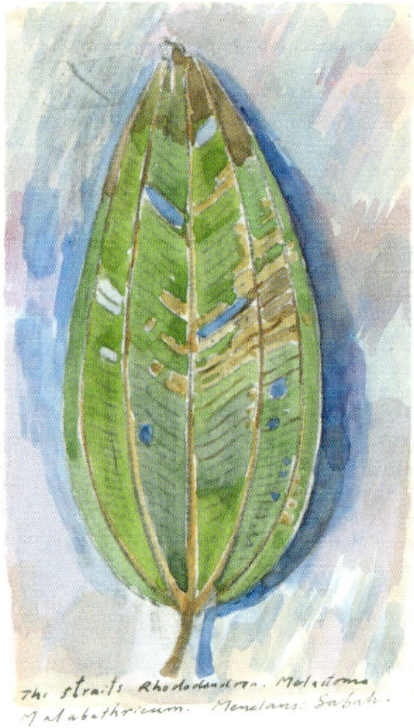

The classical side of my character (rather a large proportion) wants to see that these growth structures have evolved as the most effective way of organising space in competition or in association with all the other life forms. Thus in forests there is the urge to move to the sun (light) and create a structure which is a sculptural design to maximise this urge. The astonishing beauty of a forest canopy is to do with the different strategies each plant and tree uses to make its universe of branches and twigs and thus create an overall form within which it opens out its buds and leaves and flowers - a kind of ecstatic poem.

Today I looked down on the crowns of magnolia starred with poppylike petals each secure on a twig held by huge fan-like branches. After a time when I draw these things I am taken over by the rhythms and balances which are part of the form-making process. And these go beyond being just an effective use of space to become one with the dance which connects them and us to the universe.

Within the structures of the great branches which enclose different worlds, the leaves and buds and flowers create their own methods of receiving and listening to the sun - the choreography of hundreds of them positioning themselves in the available space is rhythmic and reminds me of notes in a musical score. What I have for ages been puzzling over is the way each one of these forms - leaf to tree crown - has a particular individual form and thus outline. It is as if it is fulfilling - or effectively filling out its available space yet the form of silhouette is always within the characteristic outline of its species. It doesn't just occupy the available space it somehow fills it out in the most poetic way, - like that *Parkia Javanica* or the ghost footprint tree in Taman Negara whose scientific name I don't know. These organisms always seem to find the most effective way to make their leaf prints on the available meniscus of sun filled air - the greatest number of footprints on the sky.

Opposite: *Green Mountain leaf, lacewing* (detail) 2000 (trace No.4)

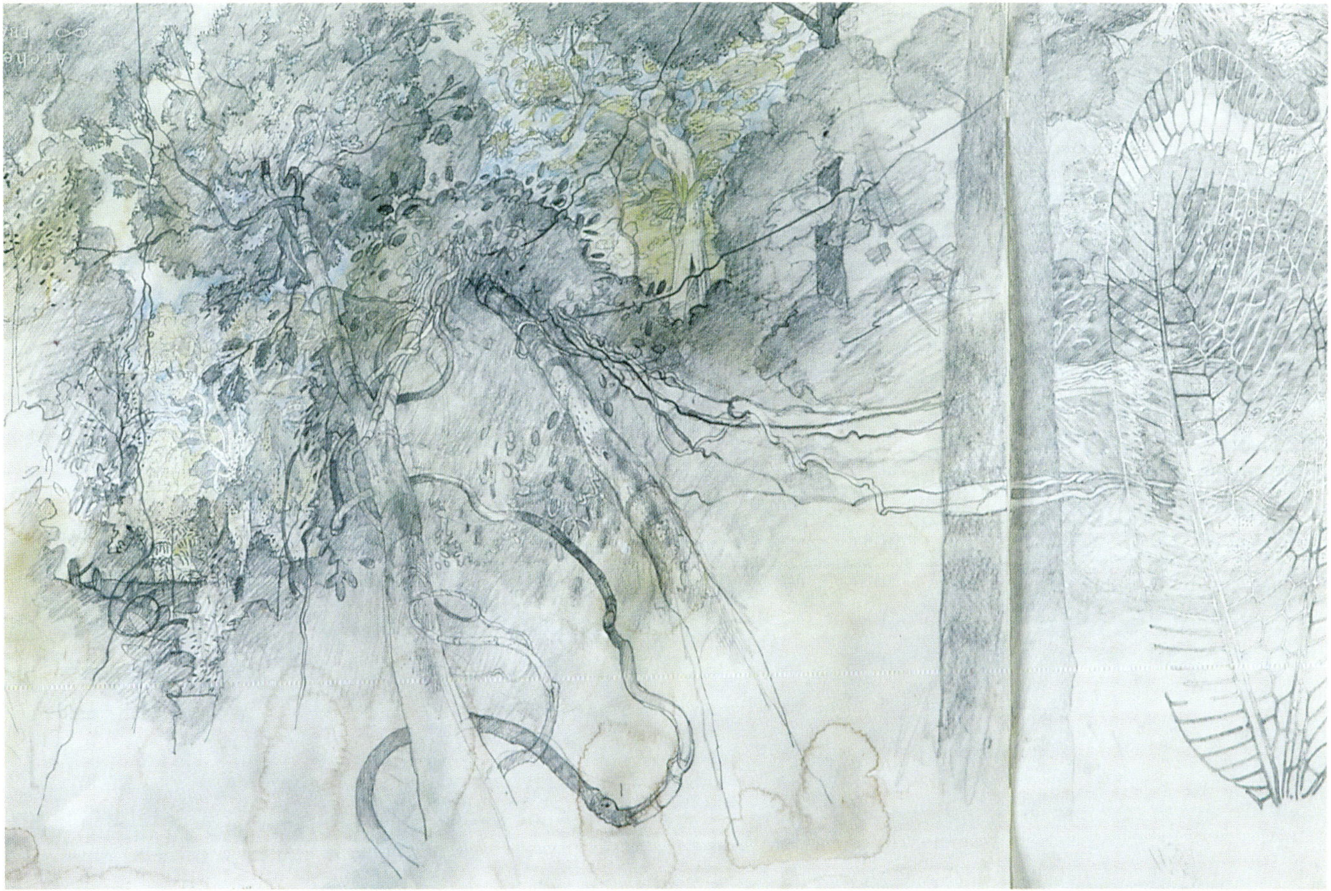

List of Works

All works are courtesy of the artist & Roslyn Oxley9 Gallery, Sydney.

TRACE NO.

1 *Javan leaf and Tiwi leaf, continental drift* 2001
watercolour on paper 56.0 x 199.0cm

2 *Rain forest canopy, olive backed oriole* 1999-2000
watercolour on paper 56.0 x 271.0cm

3 *Bower bird, the space between trees* 1999
graphite on paper 56.0 x 152.0cm

4 *Green Mountain, leaf, lacewing* 2000
graphite, watercolour on paper, two panels
56.0 x 111.0 and 56.0 x 41.0cm

5 *The territories of birds remembering Gondwana and Laurasia* 2001
watercolour on paper 76.0 x 263.0cm

6 *Tracing the Wallace Line: wing leaf and land* 1999
watercolour on paper 136.0 x 205.0cm

7 *Jungle tree, Parkia Javanica, Taman Negara* 2000
watercolour on paper 57.0 x 152.0cm

8 *Lichens and liverworts on a leaf, Mt Kinabalu* 2001
watercolour on paper 57.0 x 76.0cm

9 *Nepenthes Rajah* 2001
watercolour on paper 76.5 x 57.0cm

10 *Tracing the Wallace Line: descent of dipterocarp* 1999
watercolour on paper 56.0 x 190.0cm

11 *The architecture of leaves in Java and Malaysia* 2000
watercolour, pencil on paper 75.0 x 260.5cm

12 *A history of ferns with hoopoe and hooded pitohui* 2000
watercolour on paper 46.0 x 147.0cm

13 Logged jungle Sabah 2001
graphite on invoice slips 73.0 x 133.0cm

14 *Macaranga drift* 2001
watercolour on paper 76.5 x 114.0cm

15 *Tapir, Taman Negara* 2000
watercolour on paper 23.0 x 76.0cm

16 *Where Tembeling and Tahan rivers meet* 2000
watercolour on paper 57.0 x 76.0cm

17 *River Tembeling drawing* 2000
ink on paper 57.0 x 76.0cm

18 *Eighty-five leaves from different sides of the Wallace Line* 1997 - 2001
With Rachel Waldock
ink , graphite, watercolour, frottage, photocopies,
dimensions variable

19 *Elia Brown, Tasmania* 1996-2000
watercolour on paper 39.0 x 60.0cm

20 *The Great Tectonic Arc: concerning the collision of the Laurasian and Gondwanan Metacontinents and the idea of the Wallace Line* 1995-2001
arc on watercolour, pencil drawings on paper, plant specimens, photographs
dimensions variable

21-22 *Reliquary Cabinets for disappearing species* 2001
2 cabinets in various woods, metal containers, rock, leaves, fur, watercolour, graphite on paper
Made by Linda Fredheim Furniture Design, Hobart

23-24 *Herbarium for the end of the Millennium, specimen, blueprint, facsimile* 1996
cabinets in various woods, drawings, watercolours
Made by Linda Fredheim Furniture Design, Hobart

Biographical Notes

John Wolseley was born in England in 1938. He lived, worked and exhibited in Paris, London and Somerset before settling in Australia in 1976. His work has always been a reflection of the particular land forms in which he has found himself, and especially the Central Australian desert, the Kimberley region and the Tasmanian highlands. His work is represented in most state galleries and numerous public and private collections. A monograph: *John Wolseley Land Marks* by Sasha Grishin was published in 1998 by Craftsman House. John Wolseley is represented by Roslyn Oxley9 Gallery in Sydney and Australian Galleries in Melbourne

Paul Carter is a writer interested in cultures of seeing. He is currently working on a book *Repressed spaces, the poetics of agoraphobia*.

Linda Fredheim was born in Tasmania, and studied at the University of Tasmania, graduating in furniture design in 1992. She is based in Hobart and has exhibited widely both in Australia and internationally.

Acknowledgements

John Wolseley would like to acknowledge and thank Karen Quinlan, Director and Tracy Cooper, Senior Curator for their support and encouragement. And also in Australia, Tony Ellwood, Felicity Green, Richmond Clendinnan, Tina Hayes, John Kean, the staff of the entomology department, Melbourne Museum, Dr Gintaras Kantvilas, Terence Bogue, John Pandolfini, Stuart Houghton, Aaron Spurr. In Irian Jaya, Denny Yomaki, Amos Yikwa. In Bali, Linda Garland. In Sabah, Henry Anson Gunsalam. In West Malaysia, Angela Hijas. In Java, Dr Harry Wiriadinata and the staff of the Herbarium Bogoriensis. In the U.K. Pat Wolseley, Dr Christopher Humphries and Rachel Waldock.

John Wolseley's exploration of the Wallace Line from 1998/2000 was made possible by a New Work Grant from the Visual Arts Crafts Fund of the Australia Council. The artist gratefully acknowledges this assistance.

Australia Council for the Arts

All photography by Terence Bogue except trace no 15

Exhibition Sponsors:

Max Ellis Signs · Bendigo Advertiser · ten victoria · 3BO fm 93.5

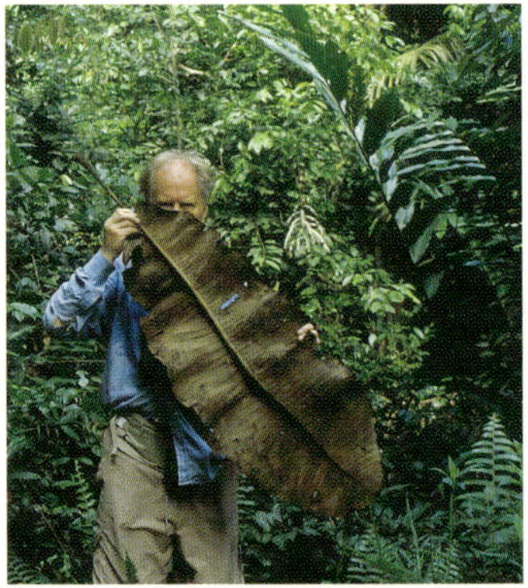

John Wolseley: Tracing the Wallace Line
Bendigo Art Gallery 7 July - 12 August 2001

National Library of Australia cataloguing in publication data: ISBN 0949215 30 9
This catalogue was printed by
Print Creations, edition 2000, July 2001

BENDIGO ART GALLERY
42 View Street Bendigo Victoria 3550
Telephone 03 5443 4991 Fax 03 5443 6586
Email bendigoartgallery@bendigo.vic.gov.au
Website: www.bendigoartgallery.com.au